Climate Change Mitigation and the European Union

A Lacanian Exploration of Desire and Enjoyment

This book focuses on the European Union's (EU) climate change mitigation actions, developing a critical framework for the 2030 Clean energy for all Europeans package and the 2050 long-term decarbonisation strategy, which informed the European Green Deal.

Reflecting on the possibility of real change and transition, the author develops a Lacanian-based discourse analysis which reflects on agentic capacities and assesses both the status quo and changes in the discourse. Informed by extensive fieldwork in Brussels and at the UN Framework Convention on Climate Change Conferences of the Parties, this interdisciplinary book spans global environmental politics, international relations, EU studies, and discourse theories/analysis.

It will be of particular interest to those working in critical theory, subjectivity, language, and materiality and intersects with a wide range of disciplines such as philosophy, sociology, anthropology, geography, linguistics, and psychology.

Valeria Tolis is Lecturer in Climate Change Governance at the University of Leeds, UK. As a scholar of critical environmental politics, her research contributes to investigating the mechanisms that govern status quo and change in the production of environmental knowledge and in the formulation of environmental policies.

Interventions

The *Interventions* Series provides a globally recognised forum for high quality, innovative, and interdisciplinary research in international politics. In 15 years, we have published 150 volumes authored or edited by a diverse network of leading scholars across all career stages.

We aim to advance understanding of the key areas in which scholars working with critical, post-structural, feminist, postcolonial, decolonial, psychoanalytic, and cultural approaches have chosen to make their interventions, and to present original analyses of politically significant topics.

All titles in the Series engage with critical thinkers in philosophy, sociology, geography, politics, and other disciplines, and provide situated historical, empirical, and textual studies in international politics.

This combination of theoretically-informed, empirically-grounded work is a hallmark of the Series, which continues to shape key debates across arts, humanities, and social sciences.

We warmly invite proposals for a variety of books from both established and up-and-coming authors including: single-authored/edited survey/textbooks; 'big idea' research monographs; edited books on cutting edge topics; and the very best doctoral theses converted into research monographs.

We are very happy to discuss your ideas at any stage of the project: please contact us for advice or proposal guidelines.

Proposals should be submitted directly to the Series editors:

- Jenny Edkins (jennyedkins@hotmail.com) and
- Nick Vaughan-Williams (n.vaughan-williams@bham.ac.uk).

'As Michel Foucault has famously stated, "knowledge is not made for understanding; it is made for cutting". In this spirit The Edkins – Vaughan-Williams Interventions series solicits cutting edge, critical works that challenge mainstream understandings in international relations. It is the best place to contribute post disciplinary works that think rather than merely recognize and affirm the world recycled in IR's traditional geopolitical imaginary.'

Michael J. Shapiro, University of Hawai'i at Manoa, USA
Edited by Jenny Edkins, Aberystwyth University and
Nick Vaughan-Williams, University of Warwick

Making Refugees' Political Agency Visible
Practices of the Subject
Amelie Harbisch

Climate Change Mitigation and the European Union
A Lacanian Exploration of Desire and Enjoyment
Valeria Tolis

For more information about this series, please visit: www.routledge.com/series/INT

Climate Change Mitigation and the European Union

A Lacanian Exploration of Desire and Enjoyment

Valeria Tolis

LONDON AND NEW YORK

First published 2025
by Routledge
4 Park Square, Milton Park, Abingdon, Oxon OX14 4RN

and by Routledge
605 Third Avenue, New York, NY 10158

Routledge is an imprint of the Taylor & Francis Group, an Informa business

British Library Cataloguing-in-Publication Data
A catalogue record for this book is available from the British Library

ISBN: 978-1-032-45677-5 (hbk)
ISBN: 978-1-032-45681-2 (pbk)
ISBN: 978-1-003-37822-8 (ebk)

DOI: 10.4324/9781003378228

Typeset in Times New Roman
by Apex CoVantage, LLC

Contents

Acknowledgements

I would like to thank the editorial team for their helpful suggestions on previous drafts of this book. I am immensely grateful to Chris and to my family for their love, support, and for putting up with me during this long writing journey.

Figures

Tables

1   Introduction

The anthropogenic cause of global warming resulting from the continuous emissions of greenhouse gases (GHGs) into the atmosphere is no longer contested within the scientific community. Cyclical scientific reports from the IPCC, the United Nations Intergovernmental Panel on Climate change (2023), constantly reinforce the need for ambitious climate change mitigation actions as well as 'an urgent and fundamental departure from business as usual' (IPCC AR5, 2014, v) to limit global warming to below 2°C and possibly 1.5°C compared to pre-industrial levels (IPCC AR5, 2014 and IPCC special report 2018, 51). Today, locutions such as the 'climate crisis', the 'climate emergency', the 'climate challenge', or the 'climate breakdown' have become normalised in the socio-political discourse and entered the everyday vocabulary across all societal sectors and governments, and national pledges and plans for 'decarbonising the economy' or achieving 'net zero' by 2050 or 2060 – including from big emitters such as China and the US – have spread across the globe.

The perceived green reputation of the EU as its overall consistent involvement with environmental policy since the 1970s constitutes the starting point of this discussion. The EU – historically among the Western big emitters – has traditionally been responsive to scientifically informed climate warnings (IPCC, 2023) and has pushed for binding international commitments since the Kyoto Protocol (1997) under the United Nations Framework Convention on Climate Change (UNFCCC). Although its domestic climate mitigation policy has proceeded in fits and starts throughout the 1990s and early 2010s, an intense policy activity has characterised the period between 2015 and 2019, intertwining with the ratification of the Paris Agreement (2015),[1] and culminated with the launch of the European Green Deal, an overarching plan aiming to deliver decarbonisation and climate neutrality by 2050 (European Commission, 2019).

This book aims to investigate the nature of the EU's climate action and on what constitutes 'change' in relation to both the logic of fundamental departure from business as usual and its ambitious mitigation policies. It contributes to discussions about change and transformation within the field of International Relations (IR) by asking the following questions: is it possible to bring about a system change? How can actors disrupt and possibly subvert the status quo and what, conversely, prevents change? Finally, how can we investigate change and distinguish real

DOI: 10.4324/9781003378228-1

transformation from an apparent and fictitious change? To answer these questions, I will contribute to the poststructuralist strand in the wider global environmental politics (GEP) research sub-strand of IR and more widely to critical IR theory by building on Jacques Lacan's theory of discourse and theory of the subject. At the most basic level, a Lacanian perspective makes it possible to place a great emphasis on language and signification, while at the same time retaining the centrality of the subject by virtue of a dialectical ontology emphasising a constitutive lack in the socio-linguistic – that is to say, discursive – production of subjectivity. This approach makes it possible to analyse the effects of the signifying chain on the speaking subjects, detect the fractures of the discourse, and assess the coping mechanisms of the speaking subjects in dealing with the failures of the discourse, which ultimately link back to Lacan's critique of how change happens via his theory of the four discourses.

Why a Lacanian reading of environmental politics?

In general, 'climate change' can be a complex object of research for two main reasons. First, 'climate change' is a bio-physical phenomenon that can be placed within the broader category of (human-induced) environmental degradation including, inter alia, pollution, and biodiversity loss. Second, 'climate change' is also a socio-economic issue, notably related to problems of energy (in)security, fight over natural resources, and forced migrations. Therefore, investigating climate change always entails a risk of reductionism, that is, treating it as a single-issue area and excluding other interrelated factors in the analysis. Climate change politics is also a complex object of research due to the interplay of a wide range of actors and institutions as well as to the multi-level character of global climate governance, which inevitably affects the choice of the starting level of the analysis.

As a result, in this book, I will emphasise the socio-linguistic performative character of 'climate change' and consider 'mitigation' as the starting point of reflection from a qualitative perspective, through an investigation of *how* the very concept of mitigation is thought and constituted through its speaking actors. As opposed to positivists who treat language as a closed system of ready-made tools to convey meanings coming from the outside, common to the so-called 'poststructuralists' is the view that language itself is a field of social practice. While the poststructuralist strand and discourse analysis within IR and environmental politics are heterogeneous and interconnected strands of research encompassing a vast array of approaches which address the relationship between language, discourse, and the social world (Leipold et al., 2019), discourse is overall regarded as a system or structure of signification that constructs and shapes social reality (Milliken, 1999), where meaning is constituted and contingent.[2] As the discourse provides a framework for understanding the world, it constitutes power relations as it enables and constrains what can be thought, it embeds and excludes, defines who is allowed to speak and thus produces objects and subjects (Doty, 1993; Dunn and Neumann, 2016, 47–54; Hansen, 2006; Milliken, 1999, 229; Oels, 2005). As a result, focusing on the discourse does not presume core properties or 'essences' of both objects and subjects and does not engage with assumptions about the internal structure of identities, but maintains that these cannot be grasped outside the language used to describe them (Solomon, 2015, 12). The appeal of a discursive approach for environmental scholars lies first in how it enabled them to emphasise the socio-linguistic performative character of 'climate change'. For example, the available scholarship highlighted that the concepts that populate the language of environmental politics such as 'nature' (Beck, 1995; Haraway, 1991), 'environment' or 'climate' are continuously (re)invented and contested in a meaning struggle regarding their interpretation and implementation through environmental policy-making, planning, research, and development as well as everyday practices (Feindt and Oels, 2005; Hajer and Versteeg, 2005, 176). Moreover, a discursive approach shed a light on the complexity of actors and institutions that populate the environmental policy process (Richardson and Sharp, 2001, 194), which becomes particularly challenging to map when dealing with multi-level and/or multi-stakeholder governance. For example, in the case of climate change – as a global challenge requiring concerted action at multiple jurisdictional and territorial levels – a system of governance has evolved since the 1992 United Nations Framework Convention on Climate Change and became more complex, resulting in an intricate network of state and non-state governance. Therefore, when we choose a political actor, we are confronted with the dilemma of which unit of analysis to choose or start from. For example, in the case of the EU, we can choose whether to include (or exclude) the EU's international role in the (international) climate governance of climate change, or we could focus on the EU institutional level, and it is also possible to extend the analysis to the level of the Member States or even at the level of regions and/or cities insofar as the decisions taken at institutional level impact the Member States. As the discourse does not assume an individual or a state-self, it starts from a basic question, 'Who speaks?' and this starting point makes it possible to travel from the individual to the institutional level without pre-assuming who these actors are (Epstein, 2011, 342).

Further, a focus on the discourse allows us to grasp how language is ingrained in practices and the material because the discourse operates by defining knowledgeable practices and consequently defines and constrains the range of policy options (Doty, 1993; Griffin, 2009; Hansen, 2006; Herschinger, 2011; Milliken, 1999; Oels, 2005; Shepherd, 2013; Weber, 1998). By enabling and constraining the range of the 'sayable', the 'thinkable', and the 'doable', discursive practices construct 'knowledge' – that is, ways of knowing a given subject or a given object – and exert power relations by including and excluding these objects and subjects.

Drawing on the work of Michel Foucault on disciplinary power,[3] regimes of truth and governmentality (Foucault, 1972, 1977, 1979, 1980) – where power relations and knowledge production are discursive – a focus on the discourse made it possible to answer the 'how' questions – that is, illuminate why some policies come about based on the definitions and the truths produced in given time and space (Epstein, 2011; Hajer and Versteeg, 2005, 177). For instance, discursive approaches called into question whether policy-making is about nature and climate or about those power relations produced and perpetrated in the name of the environment (Feindt and Oels, 2005, 163). Similarly, they demonstrated how the contradictions and uncertainties of science can be manipulated and exploited by politics in the case of the ozone depletion leading to the Montreal Protocol (Litfin, 1994) as well as how policy-making is a meaningful struggle between competing discourses which continuously re-define and re-frame solutions to environmental issues (Hajer, 1995).[4] Finally, these approaches shed a light on how new subject positions are produced due to environmental activism – for example, from whaling state to anti-whiling state (Epstein, 2008) – as well as how different types of 'climate change' are produced because of shifts in modes of governance from bio-power to advanced liberal government in Western industrialised countries (Oels, 2005).

Over time the term 'discourse' has been understood and operationalised in different ways, notably in relation to the relationship between language and materiality, which remains an open debate in Western philosophy of science. More to the point, some discourse scholars such as Fairclough's (2010) and Wodak's (2011) retained the dichotomy by treating the discursive and non-discursive separately and rejected 'discourse imperialism' (Fairclough, 2010, 165–166),[5] whereas others argued that social practices and materiality cannot be grasped outside the discourse as it is through language and the system of signification that objects, subjects, states material structures are endowed with meaning and identity (Dunn and Neumann, 2016, 43–46, 62; Hansen, 2006, 16–20; Laclau and Mouffe, 1985, 108; Milliken, 1999).[6] At the same time, this perceived excessive emphasis on language and (the politics of) representation exposed poststructuralism and discourse studies to the criticism of more recent theoretical frameworks, which questioned the extent to which poststructuralism in IR is able to account for action and materiality in contemporary political life. For example, the alleged neglect of action and practices spurred the emergence of a sociology-inspired practice turn (Adler-Nissen and Pouliot, 2011; Hughes, 2015; Neumann, 2002), which emphasised practices and processes in the making of world politics, from multilateral diplomacy to environmental negotiations. Another example is represented by a new materialist turn, which shifted the focus to the constitutive and performative role of objects and materiality (Bennett, 2010; Coole, 2013). In this respect, Lundborg and Vaughan-Williams (2015) observed that although new materialism deserves credit for reminding us of how 'things' and the material shape social and political interactions, behaviours, and outcomes (Lundborg and Vaughan-Williams, 2015, 12), it failed to appreciate the role of language in human/non-human assemblages.[7] At the same time, they argued that the separation between the ideational (language) and the material – where a discourse becomes a proxy for language and is privileged over the material – pertains in fact to some secondary poststructuralist work (see Campbell, 1992; Hansen, 2006; Milliken, 1999), whereas early and foundational poststructuralist work, which directly built on the work of Foucault and Derrida (see Connolly, 1974; Der Derian, 1987; Shapiro, 1981), treated materiality as fundamentally inseparable from the ideational (Lundborg and Vaughan-Williams, 2015, 11–18). In fact, for Foucault discourse is more than a collection of utterances, it is also a set of practices which produce the objects and subjects of which these utterances speak. What can be thought and said about for example, 'mad' people, 'homosexuals', and 'doctors' is defined and prescribed by that discourse. For example, 'madness' is produced in a network of both linguistic and non-linguistic relations, such as social institutions, scientific statements, as well as art and visual representations. Likewise in Discipline and Punish (1977), disciplinary power emerges with the practices that render crime and the criminal visible: for instance, the prison is the place which constitutes both the identity of society as well as the prisoner inside, in the separation and constitution of what is good and civilised through the bad and barbaric (Dunn and Neumann, 2016, 44; Hansen, 2006, 16; Shapiro, 1981, 218). In other words, practices of coercion and surveillance of bodies through punishment techniques are made possible by categorising criminals and by imposing a certain gaze that makes these bodies visible via control and surveillance, yet that body is produced out of a specific way of articulating and looking at the social world (Lundborg and Vaughan-Williams, 2015, 19; Kapoor, 2020, 49).

As we shall see, Jacques Lacan's understanding of discourse maintains the fundamental inseparability between language and materiality. As Kapoor puts it, by saying that we are linguistic beings, Lacan is not advancing an idealist ontology and saying that we create the material world, but only that our signifying chains frame reality establishing the boundaries and the coordinates for understanding and interpreting it (Kapoor, 2020, 6). This means that any meaning-making or agentic capacity of materiality cannot be attributed to matter alone insofar as it presupposes human-speaking subjects to be maintained and perpetrated. For example, a car is a material object whose material and social (re)production establishes and perpetrates specific relationships between stakeholders, the mining industry, the car industry, and the citizens as buyers-consumers. Likewise, an EU piece of legislation such as the Regulation which bans the sale of new cars and vans with combustion engines as of 2035 (Regulation (EU) 2023/851) will inevitably reconfigure these societal relations with new critical raw materials, new consumers of electric vehicles or public transport. Similarly, EU legislative acts such as the Energy Efficiency Directive (2012/27/EU, (EU)2018/2002), or the Renewable Energy Directive (2009/28/EC. 2018/2001/EU) or policy outcomes such as the Circular Economy package and their implementation – which I will address extensively in my empirical chapters – are discursive and material implications of a kind of knowledge that is put to work and made possible by the logical boundaries of the discourse. This leads us to a sort of 'dialectical materialism' for which materiality cannot be understood without immateriality – that is, the ideational and symbolic – and one cannot make sense without the other (Žižek, 2013).

A new political subject

Although the achievements of Foucauldian discursive approaches in environmental politics – and in IR more widely – cannot be disregarded, scholars agree on the fact that these downplay the role of the subject, as the latter is constrained and subjugated to power, which in turn hinders the possibility of resistance and change (Bracher and Alcorn, 1994, 29–35; Epstein, 2011, 338; see also Adler, 1997; Butler, 1997). In other words, although Foucault provided a radical analysis of power as decentred, diffuse, and relational constituting all social identities,[8] if such power permeates society and operates in the subject without any mediation or internalisation process (Copjec, 1994; Kapoor, 2020), there can be no easy identifiable target for a revolution. As a result, the distinction between society and power ceases to exist and there is no such thing as a privileged subject, emancipated from power and an essential position of resistance and subversion beyond that power cannot exist (Newman, 2004, 140–145). Similarly, Kapoor maintained that Foucault's theory cannot account for how a subject can apprehend power nor how social orders endure or can be subverted precisely because such conceptualisation does not contemplate any gap between discourse qua societal structure and its 'positive content or effect', which makes it difficult to explain their role as analyst of discourse or how the subject can make social change happen (Kapoor, 2020, 36–37). The disappearing of the subject is demonstrated in IR-GEP by the fact that except for Litfin who, based on her interpretation of Foucault, claimed that we should not be misled by Foucault's rhetoric in that his theories need subjectivity, as actors do not act as autonomous, but in a contingent subjectivity (Litfin, 1994, 23), other scholars felt the need to integrate Foucault's theory of discourse. For instance, Hajer complemented Foucault's discourse with the theoretical insights provided by two social psychologists, that is, Harré and Billig, who – in Hajer's view – complement Foucault's discourse theory given their focus on social interactions in actual speech situations (Hajer, 1995, 52–53). Similarly, Epstein (2008) made use of a mixed theoretical framework, including Pierre Bourdieu's practice theory, to illustrate different aspects of the anti-whaling discourse.

Before sketching out the Lacanian concepts used to investigate the EU's climate mitigation action, one main aspect must be clarified. Jacques Lacan is known for his psychoanalytic theory and practice; however, it is beyond the scope of this book to outline his entire clinical work. In fact, this book focuses on his theory of discourse and therefore builds on his late contributions between the mid-1960s and early 1970s, where his reflections shifted towards a more explicit social critique. This phase, among other things, demonstrates how Lacan was inevitably immersed in the cultural environment of that time, where the importance of discourse was highly explored in French philosophy, as it can be seen in the lifework of Michel Foucault, Jacques Derrida, and Gilles Deleuze.

Nevertheless, it is impossible to speak of the Lacanian notion of discourse without addressing the centrality of language in his overall theory of subjectivity, as it was developed in the earlier stages of his research and professional activity. Therefore, while a Lacanian-based discursive power-knowledge relationship is in line with that of Foucault and its applications in IR-GEP, the novel aspect of Lacan is that he retains the central role of the subject.

In Lacanian terms, 'discourse' refers to the social link founded on language, that is, an overarching socio-linguistic structure that defines and regulates our societal lives inter-subjective relations as speaking beings. When we adopt a Lacanian perspective, we are spoken by language, yet the Lacanian discourse-subject relationship can be understood by considering the paradox of language. As I shall explain more in-depth in Chapter 3, the socio-linguistic order can never fully complete this subject, thus the activity of speaking always produces an excess of discourse, a surplus of meaning, which is perceived like a loss in the enunciation. The result is that this mechanism triggers the subject's desire for totality with language and representation, but the full and conclusive meaning is always deferred, and discourse always appears open, fractured, and inconsistent.[9] In other words, resorting to an existing signifying structure to communicate introduces lack and thus also the desire to fill this lack, which, in the final analysis, undermines any sense of fullness we might seek. Discourse and subject are therefore determined dialectically and from this perspective, we cannot draw a line of separation between the inside of the subject and the outside of the discursive world (Tomšič and Zevnik, 2016, 5). In other words, discourse and subjects are two different entities, but cannot be clearly separated, and presuppose each other. Subjectivity, therefore, is lacking in its, by definition, incomplete structural relation with language, but at the same time, the social linguistic structure is itself partial and lacking too as it cannot complete the subject. In truth, nothing is really opposed to language, to its symbolic world, which Lacan calls the Other (Lacan Seminar VI): in itself, it is void (Pavon Cuellar et al., 2010, 257), a fiction sustained dialectically by itself and by the socialised subject.

As a result, Lacan's psychoanalytic theory provides a useful framework for social and political research because it recasts the IR structure-agency debate dialectically by drawing on Hegel's and Marx's dialectical ontology. More to the point, the relevance of Lacan's theory of social and political analysis emerges when we consider the socio-political production of subjectivity and the refusal of any essentialist and simplistic definition of subjectivity (Epstein, 2011, 337–338; Stavrakakis, 1999, 14). In contrast with the view that in a globalised world there are no strong social links 'and that there are only more or less dispersed individuals able to make decisions and free choices' (Klepec, 2016, 116), a Lacanian perspective points to the fact that we cannot function without a discourse, that is without a socio-political signifying structure that establishes and perpetrates specific socio-symbolic relationships. More importantly, Lacan's theory emphasises the socio-political production of an incomplete subjectivity, as the subject attempts to fill this lack through available social discursive representations providing them with a stable – yet by definition ambiguous and fragile – identity. Hence, what happens at the level of the individual is a productive circular play of lack and identification of the level of representation: the subject constantly attempts to cover this lack through symbolic identification and via socio-linguistic order that precedes them (Stavrakakis, 1999, 28–36). Instead, by emphasising a constitutive lack rather than the essence of the individual psyche and by conflating dialectically subjectivity and lack in their relation to a socio-linguistic order, Lacan avoids the essentialist reductionism of the social to the individual level, while also avoiding postmodern relativism, as well as 'the artificial application of psychoanalysis from the individual to the collective level' (Mandelbaum, 2022, 12).

The Lacanian subject thus subverted the Kantian subject that dominated agency debates in Western philosophy, which constitutes the foundation of science – including most social and political research – and permeates all our concepts and practices of being, knowledge, and conduct (Burgess, 2017, 659; Zevnik, 2016).[10]

More specifically, the Lacanian subject is not the un-divided, conscious, and self-sustaining actor endowed with a fixed identity, agency, and free will who is more or less able to act rationally and autonomously on processes and structures, as most rationalist approaches would claim. Rather, the Lacanian subject experiences the world in a different way from the Kantian subject because they are driven by desire (introduced by lack) and not reason: while the Kantian subject applies reason to a graspable reality, the Lacanian subject 'experiences a world in which the fullness of its object is elusive, where the real is only detectable as a suspicion of reality' (Burgess, 2017, 660).

This ontological shift makes it possible to configure a different place of power, meant as 'an abstract symbolic position, through which both power relations and political identities are organised and constituted' (Newman, 2004, 140). For Newman (2004), politics involves a contamination of both a universal dimension and a particular one, with identities being split between the universal qua common background – that is, the discourse – and their own particularity – their own (split) identities. Within this framework, the universal permeates the particular so that particular identities cannot exist outside a dimension of universality that permeates them. This means that a group cannot purely assert its particularity because that particular or differential identity is constituted in relation to other groups (see Laclau, 1996, 48). For example, the demands of a political or cultural minority group are always constituted through a universal dimension, insofar as the demand for the right to be different is at the same time a demand for equal rights with other groups. At the same time, the universal is also permeated by the particular, to the extent that the universal is empty (a necessary fiction) and can only represent and articulate itself through a particular identity, even if no identity can fully wholly embody this universal. This makes the (full) universal an impossible object which makes representation impossible but necessary. In this respect, 'society' is an impossible object whose universal dimension can be made representable though a particular ideology or particular political identity, such as communism, for instance. This means that no particularity can wholly represent this universal idea of society, but to have any conception of politics, we must have some symbolisation, albeit partial (Newman, 2004, 152). This different ontology makes it possible to conceive a different subject as well as remap antagonism, resistance, and potentially transformation because socio-political identities are in fact instable and lacking. The excess residue of discourse – that is, this dimension of lack and emptiness – is key in Lacan because this makes it possible to understand resistance and antagonism in a different way from that of Foucault. While for Foucault the political field displaces itself via the struggle of forces, antagonism, difference and multiplicity, for Lacan the political field is held together by lack qua discursive limit, and therefore open (Newman, 2004, 148). Antagonism then is not diffuse but structured around its lack, which ultimately renders political identities unstable as well as working as the limit around which these are constituted. As we shall see in the next chapters and notably in Lacan's theory of the four discourses – formalised in the two status quo discourses and the two discourses of rupture and change – the way this excess to discourse is handled by the speaking subjects is key to understand real change and transformation.

The EU's desire for climate action

In this project, I will investigate the EU's climate mitigation action by virtue of the higher level of ambition proclaimed by the EU both in the medium term (2030) and in the long term (2050). Consequently, I situate my research in the moment that corresponds to the finalisation of the 2030 medium-term climate policy framework and the launch of the 2050 'long-term strategy', the phase that immediately precedes what is today known as the EU Green Deal. For this purpose, throughout the project, the period under observation will be referred to as the period 'in-between strategies' to place an emphasis on the alleged transitional aspect that this policy-making should bring about in the achievement of climate objectives in line with climate science. While the book acknowledges any more recent updates such as the launch of the European Green Deal (2019) and the Fit for 55 Package, this Lacanian discourse analysis developed in this book is informed by an extensive fieldwork activity conducted in the period 'in-between strategies' at the Brussels headquarters and at the level of the UNFCCC Conferences of the Parties (COP), where I participated in both the COP23 (2017) and the COP24 (2018). In these field sites of climate policy-making, I followed the work of the EU and its stakeholders – via direct observations of UNFCCC sessions, stakeholder consultations within the EU Commission, and the EU Parliament working sessions – as well as conducted in-depth interviews.[11] To detect the EU's climate change mitigation enunciators, I looked for a site or multiple sites of observation. The events of which I provided an account in these empirical chapters are diverse in terms of scope and content. Some have a broad focus on climate and energy policy, such as 'the EU vision for a clean, modern and competitive economy' (see Table A.2 in Appendix, 10–11 July 2018). Others are not specifically climate-focused, such as the 'EU trade policy day', which included a session dedicated to a general idea of 'sustainability' (Table A.2 in Appendix, 27 November 2018). A few events are sectorial and focus, for example, on one specific technology, such as the 'Carbon Capture and Utilisation Technologies' (Table A.2 in Appendix, 17 September 2018). Some of these events fit a policy-making process, such as the aforementioned stakeholder consultation 'the EU vision for a clean, modern and competitive economy', whereas other events are stock-taking events, such as the 'EU raw materials week' (see Table A.2 in Appendix, 12–16 November 2018). Finally, some events are organised and run by EU institutions, such as all the aforementioned events (Tables A.1 and A.3 in Appendix), while other sessions are not directly led by the EU institutions but host instead a keynote speaker from a Commission member or include EU officials on their panels, such as the case of the Black Carbon event (Table A.2 in Appendix, 19 November 2018) and the Postgrowth Conference (Table A.2 in Appendix, 18–19 September 2018). These events are quite diverse and do not hold the same relevance in terms of final decision-making or policy outcome. For example, these 'unofficial' events – such as the Postgrowth Conference – do not have to be regarded as the official policy line but only as a platform for discussion facilitated by the European Parliament as a democratic institution. In interviewing participants in the EU Commission, a horizontal, cross-sectoral approach to the different Commission's departments – named Directorate General (DG) – involved with climate policy-making was adopted. These include DG Clima, DG Energy, DG Environment, DG Grow, DG Research and Innovation, and DG Joint Research Centre (JRC). When it was not possible to obtain an interview from a given DG or stakeholder, I instead identified and contacted actors who interact and gravitate around the Commission, such as the European Economic Social Committee, or Environmental NGOs such as Climate Action Network Europe. In the critical phase of the policy-making that sets out the future direction of work, I reflect on the nature of the current EU's climate action and on what constitutes 'change' in relation to the business-as-usual logic.

In the analysis of how the concept of climate change mitigation is thought of and constituted in the phase of policy-making under observation, my critique is not addressed to mainstream market-based policy solutions such as the famous EU Emissions Trading System (EU ETS) or to future-looking technofixes such as Carbon Capture and Storage (CCS) technologies. Rather, my analysis of the EU's climate mitigation action is addressed to policy tools and concepts through which climate change is rendered governable and which are apparently unproblematic and even desirable, such as *energy efficiency, renewables*, and *circular economy*. Hence, my selection of the case studies is justified on two grounds. First, these tools play a key role in both the 2030 medium-term strategy – which evolved from the Clean energy package to the Fit for 55 Package – and the 2050 long-term strategy. Second, these policies are in fact apparently desirable solutions in that in principle no one would argue against having equipment or devices that consume less energy, use renewable energy, or are recycled and re-manufactured and therefore can potentially be regarded as desirable change carriers. Through an empirical investigation of how *energy efficiency, renewables*, and *circular economy* have become some of the policies of choice, I illustrate what 'becoming energy efficient', 'becoming renewable', and 'becoming circular' mean for the EU in relation to delivering an ambitious climate change mitigation action. With the aim of reflecting on the status quo and change and with the help of Lacanian-informed discourse analysis, these policy tools are regarded as the formalisation of an overarching socio-linguistic structure that enables us to establish, define and maintain our societal, and thus climate, relationships – that is the discourse. By opening a seemingly closed discourse and exposing any fractures in the enunciation through the Lacanian-speaking subject, I demonstrate that we cannot think of *energy efficiency* or *renewables* or *circular economy* as a presupposition of sense that automatically delivers the desired emissions reductions, and I emphasise the partial and never-achieved character of *energy efficiency, renewables*, and *circular economy* as transition-carriers by exposing the inconsistencies, gaps, and blind spots which embody the ontological residue of signification of the Lacanian-speaking subject. The key to understanding the EU's current climate action then lies in the way any produced fractures are handled in signification, that is, by being positively integrated into signification or by disrupting it. In fact, I will show that with the exception of a few elements of disruption that leave room for a real rupture and possibly a change of paradigm, the EU's climate mitigation action under investigation appears more a fictitious change than a real transition. This argument is justified on the grounds that the traumatic points of the discourse, through which these fractures could become effective in bringing about change, are mainly positively integrated into the hegemonic signification, and turned into commodified knowledge and commodified objects of identification. The next six chapters develop this case study as follows.

Plan of the book

Chapter 2 provides a concise summary of the EU's climate action since its establishment in the early 1990s. The chapter aims to introduce the policy context to provide the reader with the necessary points of reference when the Lacanian discourse analysis of the EU's climate action will be conducted. More to the point, the chapter aims to demonstrate how the EU's climate policy-making can be placed at the crossroad between an institutional level and the UNFCCC international governance and these two processes mutually shape each other. The essentially informative character of the chapter introduced a degree of familiarity with the EU's climate polity, such as its multi-level governance character, its institutional architecture, the different areas of competence and law-making processes as well as milestone policy events and recent policy updates.

Chapter 3 sets out a conceptual framework based on Lacan's theory of discourse. The first part of the chapter elaborates on signification, discourse and subject from a Lacanian perspective, and insists on their relationship of mutual presupposition. 'Discourse' becomes then an overarching socio-linguistic structure allowing subjects to establish and organise their inter-subjective societal, political, and by extension 'climate' relations. In other words, this is the network of signification and practices those subjects presuppose to create sense, and more widely, to establish and maintain these societal relations. At the same time, discourse presupposes and necessitates a real speaking subject to be established and maintained. However – and this is key – the activity of speaking always produces an excess of meaning that is perceived like a loss in the enunciation, as the socio-linguistic order can never fully complete the subject. This theory thus places at the heart of the analysis a lacking but socially produced climate subject, that is split – by way of example – between the pursuit of climate objectives and a signifying structure defining them and regulating their actions. The second part of the chapter addresses the theory of the four discourses as these will be used as theoretical guidelines to reflect on what underpins knowledge in the EU's climate mitigation action and how we understand the 'objects' of identification that are relevant to this transition, such as *energy efficiency*, *renewables*, or *circular economy*. These discourses produce different social effects: the Master's (with its Capitalist variant) and the University discourses embody the hegemonic discourses of power and commanded knowledge, whereas the Hysteric's and the Analyst's discourses embody the discourse of challenge and ultimately change. Finally, in this chapter, I briefly explain how I turned this theoretical framework into a Lacanian discourse analysis by starting from 'the enunciating act', which is the moment when the mutually presupposed relationship between discourse and subject is manifested. I therefore started from the system of structuration of language and from the produced unfolding signify-ing chain as 'knowledge at work' in which the observed and interviewed subjects are caught, and interpreted these in conjunction with Lacan's theory of the four discourses. The impossibility of the discourse is rendered visible in the blind spots, contradictions, and weaknesses that make the discourse always partial and incon-sistent. Mapping the structuration of language based on produced spoken patterns allows us to expose the unintended effects of language on the subjects and it is the precisely this produced ontological residue of signification that explains rupture and change based on how this fracture is positively integrated into signification or how it disrupts signification.

Chapter 4 presents the starting point of analysis and investigates the ways in which the EU's climate mitigation action is thought and constituted by the speak-ing subjects. Within these spoken signifying structures, discursive interdependen-cies between climate, energy, environment, and the economy are constituted in relation to the period referred to as 'in-between strategies'. The focus of this chap-ter is to reflect on the type of 'knowledge' that underpins the EU's climate action, and the aim is to generate alternative understandings of a seemingly closed dis-course to reflect on the nature of the EU's climate transition and on any possibility of shaking the foundation of the discourse. The central element of the chapter is not so much the detection of the real authority of the social bond but rather its disavowal. Via a Lacanian-inspired discourse analysis, I show how – in the pursuit of achieving the goal of climate change mitigation – the EU's climate mitigation action is placed by policy-makers in a context of knowledge which, despite being hailed as evidence-based, is put to work in function of a disavowed command of competitiveness, accumulation, and ultimately growth. Despite this discursive clo-sure, I demonstrate how it is still possible to disrupt a seemingly closed discourse and expose its fractures. To accomplish this, I expose the nonsensical, or partial, character of *climate change* and *mitigation* and show how these signifiers are even-tually signified around the semantic gravitation centre of *energy transition*, which potentially closes the signification space by neglecting considerations about the ecosystem, biodiversity, and the planet more in general. Yet this closed significa-tion and understanding of 'knowledge' can be disrupted with the help of hysterical subjects present in the field.

Chapter 5 zooms in on this policy context of 'knowledge' and presents the cases of *energy efficiency* and *renewable*s. Through Lacanian discourse analysis, I empirically show how these policy tools – however desirable – constitute the quintessence of the rationalised and valorised commanded knowledge produced via the integration into the signification of its traumatic points. The closure of representation – mainly around the climate and energy nexus – results in turn in a paradoxical (dis)satisfaction and fictitious sense of plenitude of targets, impact assessments, continuous consultations, modelling, policy drafting, technology innovation and research, or efficient or renewable consumption. For this reason, I conclude that *energy efficiency* or *renewables* as a presupposition of sense do not automatically lead to the desired and necessary climate change mitigation. As a result, the proclaimed climate action qua energy transition as perpetrated by *energy* *efficiency* and *renewables* – and despite the hysterical action perpetrated by few actors – appears to be a fictitious change that is not able to defy the hegemonic discourse.

Chapter 6 introduces what was at that time an element of novelty in the EU's climate action, the *circular economy* – of which I will provide a 'symptomatic' reading. First, I emphasise the potentially disruptive character of the metaphor of 'circularity' – which embedded wider considerations of metabolism, complexity, and interrelation – and how this could become a common pattern across policy areas, in that it requires a consideration of flows and how these interact with the surrounding biophysical environment. More specifically, it could contribute to a greater symbiosis and coordination of all EU policy DGs, that is, the EU Commission's departments. In other words, this alternative concept would pose the circular economy knowledge in contrast with the linear, reductionist, and bureaucratised approach that has been so distinctive of the Commission. In this respect, *circular economy* is in principle acknowledged as a vital part of an effective EU's climate action. However, precisely because it pointed to a re-organisation of our societal and economic relationships and to how knowledge is produced and exchanged, the initial element of disturbance represented by 'circularity' had been gradually resisted and this resistance manifested first through a withdrawal of the whole circular economy project and then with a re-introduction and co-optation of the project.

Chapter 7 follows the case of the circular economy in its co-opted phase and highlights its features, with the main argument being that this co-optation has entailed a loss of the liberating potential that 'circularity' had previously carried with it. However, through a Lacanian discourse analysis I look at what 'becoming circular' means for the EU and again open a seemingly closed discourse to detect whether and where fractures in discourse still emerge. I conclude that *circular economy* is not spoken of as a common pattern across policy areas but has become another policy carriage of the EU qua regulatory machine. The illusion of greater circularity is sustained by the fantasies of standardising practice, increased recycling, eco-design meant as energy efficiency and new business models that perpetuate old logics, which promise to cover the impossibility of discourse and give us a semblance of circularity. Thus, the alleged circle in which the subject is caught seems to picture a loop of painful and paradoxical (dis)satisfaction which must move faster and on a larger scale without being able to picture the circularity, meant as transformation, and pursue the desired climate objectives. Yet, I also expose important elements of counter-resistance: perpetrated by stakeholders and civil society, these elements disturb the co-optation conducted by the EU and challenge the status quo, by keeping the original disruptive element of circularity alive and by expanding the circular economy signification beyond more recycling, beyond energy efficiency, beyond the ever-expanding geographical boundaries.

Chapter 8 is the conclusive chapter and in this, I summarise the contribution of this book for the way in which I investigate 'how' the very concept of 'EU's climate mitigation action' is thought and constituted through its speaking subjects.

Building on a Lacanian understanding of discourse and subject and by placing an emphasis on signification and the unintended effects of enunciation on the speaking subjects, this chapter summarises the main arguments raised in this book. These are the key themes of disavowed knowledge, the real authority of the discourse, the fractures in discourse, how the surplus of signification that emerges in the enunciation is handled, and what type of transformative forces can be detected. To accomplish this, I reassert that the partial and never-achieved ways in which signifiers such as *efficiency, renewables,* and *circular economy* manifest themselves in the moment of the enunciation should not be regarded as competing 'storylines' of 'framings', but they coexist as an ever-partial expression of the same discourse as the social link. As a result, this approach retains a more resistant subject, for its emphasis on the speaking subject as a form of agency and on its effects, that is, the produced surplus of meaning. Looking ahead, the chapter discusses their relevance by acknowledging recent developments within the framework of the EU Green Deal and broadens the scope of the book in the way it contributes to the field of environmental politics and IR (critical) theory and methodology more widely.

Notes

1 The Paris Agreement commits its parties to keep global warming below 2°C and pursue efforts to limit the increase to 1.5°C (Paris Agreement, 2015, Art. 2) but leaves to the parties how decarbonisation will be conducted.

2 Discourse analysis in political science has been influenced by different philosophical traditions (see Rasiński, 2011). The philosophical traditions have been in turn translated into different analytical frameworks. Leipold et al. (2019) include among other things: Laclau and Mouffe's (1985) Discourse Theory, Roe's (1994) Narrative Policy Analysis, Fairclough's (2010) Critical Discourse Analysis (CDA), Dryzek's (1997) Deliberative Discourse Analysis, Hajer's (1995) Argumentative Discourse Analysis (ADA), and Keller et al.'s (2018) Sociology of Knowledge Approach to Discourse (SKAD). In environmental politics, we note a predominance of socio-linguistic approaches concerned with power and ideology, for example, Fairclough's CDA (2010) as well as those attempting to merge the post-positivist approaches with a critical rationalist framework, such as the Discursive Agency Approach (DAA). Other perspectives derived from sociology and political science continue to influence environmental discourse analysis (Leipold et al., 2019, 448–449). For an in-depth review of different types of discourse analysis in social research, see also Keller, 2012; Schiffrin and Tannen, 2001).

3 Although the language turn in IR drew heavily on Foucault's work, another important strand builds on the work of Jurgen Habermas where the discourse is seen more as a process of argumentation and deliberation (Habermas, 1996) and policy as a deliberative practice (Audet, 2016; Dryzek, 1997; Fischer, 2003; Hajer and Wagenaar, 2003; Vanhulst and Beling, 2014).

4 Hajer (1995) showed how policy-making is the redefinition and reframing of solutions to environmental issues. In his case study of acid rain regulation in the UK and the Netherlands in the late 1970s, he analysed how the discourse of 'radical restructuring' had been delegitimised in both cases in favour of 'ecological modernization', on the grounds that ecological modernisation is compatible with existing political and economic structures and seeks for business opportunities in terms of innovation and the development of new markets (Hajer, 1995, 31–36).

5 According to Fairclough, the objects of analysis are both semiotic – that is to say discursive – and material (Fairclough, 2010, 206). For example, the practices of governing

can occur by law and speech acts or by resorting to violence, with the latter being non-discursive. Therefore, power in CDA is not only discursive but also material and physical (Fairclough, 2010, 4).

6 This epistemological fragmentation is not surprising if we consider that Foucault's theory of discourse has been interpreted in different ways. In analysing the conditions of appearance of an object of the discourse, Foucault pointed to the distinction between primary relations which are independent of the discourse and object of the discourse, such as relations between institutions, techniques, and social forms; secondary relations (relations in the discourse); and discursive relations which are neither internal to the discourse nor external to the discourse, but at the limit, and characterise discourse as practice (Foucault, 1972, 44–46). This would point to Foucault's himself retaining the distinction between the discursive and the non-discursive. In this respect, Rasiński (2011) as well as Dunn and Neumann (2016) highlighted that the way in which the non-discursive reality influences the discursive reality, while at the same time remaining independent, remains unclear, and it has been in fact object of debate (see Dreyfus and Rainbow, 1983; Rasiński, 2011 for in-depth methodological critique to the Archaeology of Knowledge).

7 As a result, by retaining the same distinction between language and materiality as separate and separable, it ended up reifying, rather than overcoming, the same ideational-material divide of their counterpart, but by privileging materiality over discourse (Lundborg and Vaughan-Williams, 2015, 24). For example, Drieschova (2017) suggests resorting to Peirce's semiotics as one possible way to bridge the ideational and material divide.

8 Newman (2004) emphasised that Foucault subverted the classical paradigm of the political theory of power emanating from a central and structural place. As per the foundational paradigm in classical political theory – the Hobbesian paradigm – power comes from a central place, embodied by the body of the sovereign and its temporal authority or – in modernity – by the state apparatus and its political institutions and bureaucracies. Power is also conceptualised as emanating from a central place even in opposite political philosophies, such as liberalism (where the place of power finds legitimacy in the social contract) and in radical philosophy (where the place of power is considered oppressive and illegitimate). As a result, the Manichean paradigm of classical radical politics theorised its counterpart as ontologically separated from this place of power and enables the possibility of the production of an opposite dynamic force as a resistant subject that aims at overthrowing this centre of power (e.g., through a revolution).

9 As Epstein observes, from a Lacanian perspective, as well as in all discourse theories, the process of inscription of the subject in language is not based on the acquisition of a positive, distinctly human, neurological capacity of speech as argued, for example, by Chomsky (1981) (Epstein, 2011, 336).

10 This is the type of subject that we encounter in EU climate studies which are preoccupied with the EU's international climate leadership (Gupta and Grubb, 2000; Oberthür, 2009), its level of climate 'actorness' (Bretherton and Vogler, 2006; Wurzel and Connelly, 2011) and to the extent it reduced its credibility gap between international bargaining and internal commitments (see also Jordan et al., 2010; Oberthür and Kelly, 2008; Oberthür and Pallemaerts, 2010, 12–14, 28; Skjærseth and Wettestad, 2010).

11 Tables A.1, A.2, and A.3 in Appendix provide a summary of the activities conducted. Tables A.1 and A.3 are related to UNFCCC COPs and do not list the negotiating slots, contact groups, or plenaries, informal consultations in which the EU delegates participate in the international climate governance. In fact, these focus on the implementation of the Paris agreement on a global scale and do not focus on the specific EU's climate policies.

2 The EU's climate action

Policy at a critical juncture

Introduction

This chapter provides a brief overview of the EU's climate action from its origin until present times with the aim to illustrate the actors, processes and institutions involved, and facilitate the comprehension of the policy context under analysis in this book. To understand the EU's climate policy-making, we must situate it at the crossroad between an institutional and an international process that mutually shape each other along a continuum: the EU institutional process comprises the work undertaken by the EU institutions at the European headquarters in Brussels, yet this process does not take place in isolation from the wider global climate governance and is therefore inherently linked to the UNFCCC machinery. Within these two distinct but interlinked processes spanning across three decades, from the 1990s to today, it is possible to identify periods of intense activity and setbacks, recurrent trends, and governance patterns, as well as precise events and milestones in which these two processes converge. Overall, this descriptive chapter aims to provide the reader with reference points into a complex, multi-level and multi-institutional policy landscape to which a psychoanalytical filter will be applied in the remaining chapters of this book through the help of Lacan's theory of discourse.

The EU as a climate polity: between ambition, progress, and setbacks

In the previous introductory chapter, I referred to the period under observation – namely, the finalisation of the 2030 medium-term strategy and the launch of the 2050 long-term strategy – as the period 'in-between strategies' to place an emphasis on the intended transitional character of the EU's policy mitigation response, and to frame it as a possible critical juncture. To situate this in a broader policy context and before introducing Lacan's framework, I shall illustrate how the EU's climate mitigation action must be understood at the crossroad between an institutional and an international process that mutually shape each other along a continuum in which the international treaties such as Paris Agreement as well as any EU regulatory policy packages, such as the '20–20–20 package' or the 'Fit for 55 package', represent intermediate reference points for the reader in case they get lost

DOI: 10.4324/9781003378228-2

in the discussion. Although understanding the EU policy-making as a sole top-down process from the supranational institutions to its Member States might be an over-simplification, for the purposes of this book it is sufficient to acknowledge that the EU institutional level can be understood as the work that takes place at the European headquarters in Brussels, where the main EU institutions formulate policies and negotiate regulatory frameworks that will be implemented by Member States.

The EU's climate mitigation policy spans across three decades, from the early 1990s until today, and its mitigation portfolio is considered as among the most advanced in the world, as well as a relative success if compared to the rest of developed nations and historical emitters (Dupont et al., 2023). Climate policy, however, advanced in fits and starts across these three decades and was impacted by various political and economic contingencies. More specifically, while in the 1990s progress in climate policy was limited, as exemplified by the failure in agreeing on a carbon tax proposed by the European Commission and blocked by the Member States, climate change gained political salience in the 2000s, as demonstrated by an intense period of policy activity which resulted in creation of – among other things – the EU Emissions Trading System (ETS)[1] or by the development and integration of the climate and energy portfolios. Similarly, while the early 2010s have witnessed a slowdown of climate policy activity as well as a patchy implementation of commitments due to the consequences of the 2007/2008 financial crisis (Gravey and Jordan, 2021; von Homeyer et al., 2021), the period between 2015 and 2019 has witnessed instead a resurgence in climate policy activity which culminated in the 2019 European Green Deal, an overarching plan aiming to deliver decarbonisation and climate neutrality by 2050 (European Commission, 2019).

The EU's climate action does not happen in a vacuum but is embedded in a wider global climate governance machinery which – for climate change – was established by the 1992 United Nations Framework Convention on Climate Change (UNFCCC). Within the UNFCCC process, the ratification and implementation of the Paris Agreement (2015) represent the latest formal and legal development, but it is worth noting that the EU has always pushed for binding commitments since the Kyoto Protocol negotiations (1997). Therefore, on the one hand, international commitments shape internal institutional policies: for example, the Paris Agreement resulted from the collective bargaining of 196 Parties and established a mixed-type governance with top-down and bottom-up elements. Yet, although the agreement was concluded on 12 December 2015 after a two-week intense negotiation activity, it is in fact the result of a process characterised by many fits and starts, which include the failure in producing an agreement at Copenhagen 2009, the Cancun conference 2010, the Durban Platform for Enhanced Action 2011 (COP17) calling for a legal instrument applying to all parties, and finally COP19 in Warsaw which urged parties to send their Nationally Determined Contributions (NDCs) (Ciplet et al., 2015), namely their mitigation and adaptation plans. This means that it mandated a global target (the top-down element), by requiring signatory parties to keep global warming well below 2°C and pursue efforts to limit the increase to 1.5°C from pre-industrial levels (Paris Agreement, 2015, Art. 2), but leaves to the parties to determine how decarbonisation, thus mitigation, will be conducted (the bottom-up element); accordingly, parties are required to submit their NDCs, which are revised every five years. In the UNFCCC forum, the EU – who has a legal personality – is represented by the European Commission, who takes part in formal negotiations. During the two weeks of UNFCCC COPs, the EU is also informally present through a series of side events (at its own 'EU pavilion') which do not have a direct influence on formal negotiations but are used as a showroom of policy activity or as platforms for discussions with stakeholders. As a result, the ratification of the Paris Agreement and the period between the first submission of their NDCs and their first five-year revision (2020) influenced the renewed period of intense policy activity 2015–2019 at the EU institutional level during which the EU finalised its 2030 commitments and paved the way for long-term commitments, which culminated in the launch of the European Green Deal in 2019.

On the other hand, viewing the EU's climate policy as a top-down process and UNFCCC driven is too simplistic and it would be more accurate to assert that the EU's climate action and the UNFCCC process mutually shape each other. For example, when the Paris Agreement was being negotiated at the UNFCCC COP21 in December 2015, the EU had been pushing for a post Kyoto international binding treaty since the 2009 COP15 in Copenhagen and was already working on its post-2020 climate and energy policies in the medium term (2030) and long term (2050). Finally, a potential third level of governance, the level of Member States or the sub-state level – that is regions and cities – can be detected. Yet, this book focuses on those sites of policy-making in which the EU acts, negotiates, and formulates policies as a whole.

EU climate scholars have identified a series of governance trends and patterns that can help the reader navigate the different policy milestones as well as this multi-layered policy landscape across the past three decades, with the ETS, renewables, energy efficiency, and effort-sharing approaches among Member States as the main climate governance tools (see Dupont et al., 2023 for an exhaustive summary). Common to EU (climate) scholars is the view that the EU is acknowledged as a leader by example in the international arena, but its international ambition does not often match its domestic action (Dimitrov, 2016; Oberthür and Dupont, 2021; Oberthür and Kelly, 2008; Oberthür and Pallemaerts, 2010) which is instead characterised by an implementation gap across Member States due to internal political divisions, notably along the West-East divide. For example, the 20 per cent GHGs emissions reduction targets set for the decade 2010–2020 – eventually (over) achieved – had been criticised at that time for falling short of IPCCC's recommendations (Dupont et al., 2023), that is a reduction for developed countries between 25 per cent and 40 per cent by 2020 (IPCC, 2007).

Another relevant feature is a reliance on decade-wide target setting, which are often the result of internal political bargaining and do not necessarily reflect the scientifically-recommended reductions (Dupont et al., 2023).[2] Targets are enshrined in policy packages, namely a series of legislative measures that have become more frequent and complex over time which are proposed by the European Commission and then negotiated between the Parliament and the EU Council, as per the Ordinary Legislative Procedure[3] (Hurka et al., 2021). For example, the decade 2010–2020 was regulated by the 20–20–20 policy package which was agreed in 2007 and entered into force in 2009 and set a target of reducing GHG emissions by 20 per cent, increasing the share of renewables by 20 per cent, and increasing energy efficiency by 20 per cent by 2020 compared to 1990 levels through a set of legislative tools. Meanwhile, in October 2014, the EU had agreed to set up a framework with increased targets for the following decade 2021–2030 which resulted in the 'Clean energy for all Europeans package', another set of legislative tools which was completed only in May 2019. The targets for the current decade (2021–2030) have been revised multiple times since 2019 because of contingent factors, including a change of EU leadership in the Commission (and new Parliament) as well as the Russia-Ukraine conflict which spurred a debate on decreasing fossil fuels imports from Russia. On its first formulation this package aimed to increase the share of renewables by 32 per cent, improve energy efficiency by 32.5 per cent and increase the reduction target of GHGs emissions to a 40 per cent reduction compared to 1990 levels. The package included a revision of existing legislation (which was included in the previous 20–20–20 package), such as the Energy Efficiency Directive[4] (Directive 2012/27/EU) and the Renewable Energy Directive (Directive, 2009/28/EC) as well as a change in governance compared to its predecessor, establishing a soft governance approach comparable to that of the Paris Agreement: the target is set top-down but all EU Member States must set up their own climate and energy plans for the period 2021–2030, with this mechanism being governed by the Governance Regulation that is part of the package (Regulation (EU) 2018/1999).

In the meantime, the need to strengthen the Paris Agreement commitments towards 2050 led to the formulation of long-term proposals, beyond 2030. In late November 2018 and ahead of the UNFCCC COP24 in Katowice the Commission issued a proposal, in EU jargon a Communication, the 'A Clean Planet for all – A European strategic long-term vision for a prosperous, modern, competitive and climate neutral economy' (COM (2018) 773 final)[5] followed a year later (2019) by another Communication the EU Green Deal (COM (2019) 640 final). The 'A Clean Planet for all' Communication itself did not constitute a revision of the 2030 targets, which were deemed insufficient to contribute to the Paris Agreement's temperature goals, but presented eight scenarios (COM (2018) 773 final in-depth supporting analysis, 56): some of these were compatible with either the 'well below the 2 degrees' objective or '1.5 degrees' objectives of the Paris Agreement and the Sustainable Development Goals, which ultimately demand 80–95 per cent decarbonisation (see COM (2018) 773 final, 3–5). Yet the subsequent Communication on the EU Green Deal seems to have inaugurated instead a more ambitious and disruptive phase (Eckert, 2021; Kulovesi and Oberthür, 2020) in two respects: first, its mitigation goals for the medium term and long term were translated into law with the European Climate Law (Regulation (EU) 2021/1119) – proposed in 2020 and adopted in June 2021 – with a further revision of targets from 'at least' 40 per cent by 2030 of the Clean energy package to 'at least' 55 per cent compared to 1990 levels.[6] Second, the EU Green Deal encourages further policy integration across sectors (Oberthür and von Homeyer, 2023) beyond climate and energy portfolios (Dupont, 2016) to include – inter alia – pollution reduction, biodiversity restoration and protection and the social implications of a 'just transition', and notably extends policy packages between climate policy and the land use sector (LULUCF, Land use, land-use change, and forestry). The newly revised targets therefore led in turn to a further policy package, the Fit for 55 Package (proposed by the Commission in July-December 2021) which strengthened existing legislation – including the Renewable Energy Directive and the Energy Efficiency Directive – and set a further target increase by 2030 of both the previous share of renewables in final energy consumption from 32 to 40 per cent, and energy efficiency improvements from 32.5 to 36 in final energy consumption and 39 per cent in primary energy consumption (von Homeyer et al., 2022, 127).[7] This package was further amended in May 2022 by the REPowerEU plan (European Commission, 2022) in response to the energy crisis exacerbated by the Ukraine-Russia conflict, which aims to accelerate the clean energy transition and among other things facilitate a transition away from fossil fuel imports from Russia – mainly reduce gas demands – by diversifying energy supply geographically[8] and by further increasing the targets of the Renewable Energy Directives from 40 to 46 per cent (including sub-targets for green hydrogen) and of the Energy Efficiency Directive from 9 to 13 per cent (COM(2022) 230; von Homeyer et al., 2022 final).[9]

However, despite this intensification of activity denotes the political salience of the climate issue as well as the ultimate overachievement of the 2020 targets by an estimated 29 per cent in 2021 compared to 1990 levels and against the set target of 20 per cent (EEA, 2023), its mitigation action is still considered insufficient given the scope, the scale and pace of the challenge, notably under the Paris Agreement goals (Dupont et al., 2023). Importantly, insofar as the journey from target setting to an assessment of the actual mitigating effectiveness is drawn out over several years, even decades, the mitigation effects and impact of some of the policies put in place today will have a continued impact after 2030 and can only assessed after 2030. For example, the impact of the 2035 ban on internal combustion engine will be observed only after that date, as well as the impact of the share of green fuels in aviation and maritime transport and of green hydrogen will become significant after 2030 (Von Homeyer et al., 2022, 127). Furthermore, scholars emphasised how ambition is often diluted by flexible mechanisms – the results of political bargaining. For instance, the 55 per cent target is weakened by flexible mechanisms such as offsetting nearly 3 per cent points through LULUCF credits (Gheuens and Oberthür, 2021).

In summary, the EU's climate action manifests itself as a multi-institution, multi-stakeholder, and multi-level political process in which we can detect many incremental policy changes and possible critical junctures. This book will now operate a shift in the ontology of the subject EU and its discourse of climate knowledge and will build a framework for making sense out this apparent fragmentation of knowledge practices to identify the mechanisms that regulate stability and change in its discourse, and it is to Jacques Lacan's theory of discourse that now I turn.

Conclusion

This chapter has provided an overview of the EU's climate action from its origin until present times to provide the reader with the necessary information on processes, institutions, climate policy milestones, and gain a basic understanding of the policy context under analysis in this book.

I have illustrated how, to contextualise the EU's climate action we must situate this within a complex, multi-level and multi-institutional policy landscape in which we can detect two interconnected processes that mutually shape each other: these are the EU institutional process, that is, what happens at the Brussels headquarters, and the global climate governance within the UNFCCC process. As I am preoccupied with the sites in which the EU negotiates, produces knowledge, and formulates policies as a whole, that is its supranational dimension, the Member State level of analysis is beyond the purposes of this book.

Within this multi-level and multi-institutional policy landscape EU climate scholars have highlighted alternating periods of policy activity and setbacks in which it is possible to identify some general trends such as progressive targets increase at each decade, their formalisation into policy packages containing legislative measures and some policy integration across sectors, as well as a reliance on long-established policy tools and an international high ambition that is usually softened at a domestic level due to internal political divisions among the EU Member States.

This relatively short and descriptive chapter provides the necessary foundational knowledge that will enable us to operate an ontological shift and conceptualise the EU as a political subject and as a new place of power under Lacanian lenses. In turn, this move will enable us to explore the mechanisms that regulate the production of climate mitigation knowledge and where and how the potential for a transformative discourse arises. Therefore, this book will now turn to a thorough exposition of Jacque's Lacan's theory of discourse to make it operational in the empirical sense.

Notes

1 To clarify, the EU Emissions Trading System is a market-based instrument that works according to the cap-and-trade principle. This means that a limit is set on the total amount of GHG emissions that can be emitted by industrial installations and aircraft operators, and this constitutes the 'cap'. The cap is reduced every year and expressed in emissions allowances that companies buy and/or trade with each other.

2 Deeply rooted political divisions remain and remerged within the EU Council and the European Parliament around the Green Deal and its implementation, notably along the East-West divide with the more progressive Western and Northern Member States and the climate-sceptic Eastern bloc, led by Poland (Skjærseth, 2021) who attempted to halt or slow down progress due to its reliance on coal for geopolitical, economic, and political reasons.

3 The OLP is the EU law-making process. The EU Treaties set out the division of competences in all policy areas: there are areas in which the EU had exclusive competences such as commercial policy; areas in which competences are shared between the EU and Members State, such as climate and environment where both the EU and Members can

both make policy, with EU measures confining the leeway for Member States; finally, there areas which predominantly still fall within Members States competences such as health and social policy. EU climate policies go through a process of intra-institutional and inter-institutional bargaining within and among the three key institutions representing different interests. These are the European Commission, the executive branch with the power to propose legislation and which represents the EU-wide position and interests; the European Parliament which is elected by and represents European citizens; and the Council of the European Union, made up of the ministers of a specific policy areas and therefore representing the member states. The Commission issues a proposal for legislation, and this will be amended by both the Parliament and the Council (who negotiate and amend the proposal separately). An inter-institutional negotiation process involving the three institutions follows where – in simple words – the Parliament and the Council need to agree on each other's amendments. Once agreements are reached and formally adopted (by voting) in the Parliament and the Council, these become law and gets published in the Official Journal of the EU. The highest political authority is however represented by the European Council (heads of states and governments) who decides on high-profile things such as the EU budget. This should not be confused with the Council of the European Union (the ministers) (European Council, 2024).

4 In EU law, a regulation is a binding legislative act, which must be directly applied across the EU. A directive sets out a goal that all EU countries must achieve and has to be translated into national laws by Member States which will decide how to reach these goals (EU, 2024).

5 From a policy perspective, a Commission Communication is a summary of what the EU Commission indicates as the recommended course of action in a given policy area before actual policies are formulated and legislation is produced; conversely, a policy package such as the 20–20–20 or the Clean energy package is a series of agreed legislative measures (Curtin and Manucharyan, 2015).

6 The Climate Law also leaves room for further developments such as the 2040 GHG targets which have not been currently agreed; an emissions budget for 2030–2050 to be agreed by 2024; the establishment of a European Scientific Advisory Board on Climate Change, and a review of other EU legislation with the 2030 and 2050 climate targets (Regulation 2021/1119, Articles 3, 4, 6 and 12).

7 The Fit for 55 package also revises the Emissions Trading System Directive; the Effort Sharing Regulation setting targets for non-ETS sectors (buildings, transport, and agriculture); the Regulation setting CO_2 emission standards for cars and vans; the Regulation on LULUCF setting a net zero target for the sector, as well as the Environmental Performance of Buildings Directive and the rules governing the gas market to integrate green hydrogen (von Homeyer et al., 2022, 126).

8 This includes the import of liquefied natural gas (LNG) from other parts of the world, such as Algeria (Bouckaert and Dupont, 2022) which in fact increases the risk of carbon lock-in due to infrastructure construction as well as result in increased methane emissions from LNG (IEA, 2023).

9 It also advocated for (more than) doubling power generation from photovoltaic by 2025 and quadrupling by 2030, for mandating the instalment of solar power panels on rooftops of new buildings, simplified permitting and planning procedures for renewable energy and legal obligations to upgrade the energy performance of existing building; finally, it also advocated for a tenfold increase in biomethane production by 2030.

3 Climate mitigation action and the EU

A Lacanian matter?

Introduction

This chapter explores how Lacan's psychoanalytic theory provides a useful framework for reconceptualising the EU as a Lacanian (political) subject, for investigating the nature of its climate change mitigation action and for reflecting on what constitutes transformative potential. My aim is to introduce a degree of familiarity with abstract terms that characterise Lacan's theoretical work, such as 'signification', 'discourse', 'subject', 'lack', 'jouissance', and it is structured as follows. First, I briefly introduce Lacan's theory of signification and explain how meanings do not exist independently but are produced from the sliding of signifiers. Yet, some signifiers are more prominent than others and these anchoring points – which Lacan calls 'Master Signifiers' – fix the chain's semantic ambiguity providing the necessary 'illusion' that reality is intelligible. Discourse then becomes the anchored signifying chain which provides the subjects with an overarching socio-linguistic structure that allows them to establish and maintain societal relations. However, the socio-linguistic order can never fully complete the subject as the activity of speaking always produces an excess of meaning, perceived as a loss in the enunciation, and Lacan explains this excess of meaning through what is arguably his most original conceptual tool, the *objet petit a*, which stands for the lack that stimulates desire and embodies the impossibility of full satisfaction. The implication here is that conclusive meaning is always deferred and the search for enjoyment and plenitude results in a state of paradoxical dissatisfaction that Lacan calls *jouissance*. Importantly, *objet petit a* as (libidinal) remainder of signification carries a liberating potential that can either disrupt signification and generate the conditions for an alternative signification or be positively integrated into signification, with the effect of this remainder being rendered visible in the blind spots and contradictions that cause ruptures in the discourse and make it look partial and inconsistent. As a result, the discourse is in fact composed of three interrelated registers that are held together by the subject and manifest themselves in enunciation: these are the Symbolic (the visible socio-linguistic structure), the Imaginary (the individual's mental representation), and the Real (the realm of *objet a* and *jouissance*, what escapes language), which become of empirical relevance in the application of a Lacanian discourse analysis. I then elaborate on Lacan's theory of the four discourses, which

DOI: 10.4324/9781003378228-3

describes different social bonds based on the social effect they produce and can be regarded as theoretical guidelines for understanding the EU's climate mitigation policy. These conceptual tools are regarded as theoretical guidelines to reflect on the nature of the EU's climate transition insofar as they place at the centre of the analysis the lacking but socially produced subjects, who are split between the pursuit of climate objectives and a historicised signifying structure defining them and regulating their actions. In this respect, the opposition between the University discourse of 'commanded knowledge' and the Hysteric's discourse of 'real and impossible knowledge', can be regarded as the starting point for reflecting on what constitutes knowledge in the EU's climate mitigation action. Similarly, the parallel between the University discourse of 'commanded knowledge' and the Capitalist discourse of commodified enjoyment constitutes an interesting point of reflection as to how the impossibility of the discourse is turned into commodified knowledge and consumption objects of identification. The final part of the chapter briefly outlines how I operationalised my theoretical framework into a Lacanian discourse analysis and will explain how my fieldwork activity became the chase for the 'moment of the enunciation', which enabled me to go beyond the enunciated fact of a drafted policy document and observe the moment of the enunciation in which the elements of the discourse, that is the socio-symbolic network and the surplus of sense (*objet a* and *jouissance*), emerge. In doing so, I explain how mapping the socio-linguistic network makes it possible to trace how knowledge is presumed throughout the signifying chain and to isolate potential Master Signifiers. More importantly, I emphasise how a Lacanian discourse analysis (LDA) makes it possible to disrupt a seemingly locked representation with the aim of generating possible alternative interpretations as how these are all spoken at different times of the enunciation. It is through the disruption and disorganisation of a seemingly consistent discourse that fractures are rendered visible, as these appear as gaps, blind spots, weaknesses, and contradictions, which embody the present but 'unsymbolisable' surplus of meaning. Once the fractures are exposed, we can observe if they are positively integrated into signification or if they disrupt signification and pave the way to produce alternative significations, and this ultimately captures if we are potentially witnessing a change of discourse.

Jacques Lacan's theory of discourse

The priority of the signifier

Lacan's theory of subjectivity cannot disregard the centrality of language, which he drew on Ferdinand De Saussure's theory of linguistic signs. According to De Saussure (1959), the constitution of the linguistic sign is a two-part process, called signification: the signifier is the word in its pure form (the mark or sound). For example, climate change as a sequence of letters or sounds c-l-i-m-a-t-e-c-h-a-n-g-e is a signifier and language is thus a sequence of signifiers; if the signifier is the form, the signified is the content, that is the meaning (or meanings) of that signifier. As another example, in the case of climate change, one of the meanings could be 'the excessive warming of our planet due to anthropogenic greenhouse emissions'. For De Saussure, the relation between signifier and signified in the process of signification is that of two equivalent levels (De Saussure, 1959), meaning that form and content exist independently. More specifically, the relationship between the two, that is what links c-l-i-m-a-t-e-c-h-a-n-g-e to 'the excessive warming of our planet due to anthropogenic greenhouse emissions', is arbitrary in the sense that there is not a natural connection between the two. This view of language is thus opposed to an expressionist one, where concepts exist in a pre-verbal state before being expressed through the material element of language, that is, the signifier (Evans, 1996, 189). In fact, De Saussure's account of this relationship has been subjected to critique (see Stavrakakis, 1999, 22–26), and Lacan himself departed from it by affirming the priority of the signifier over the signified. To be precise, this priority of the signifier must be intended as logical rather than chronological (Evans, 1996, 189), which means that the signified is not merely eliminated, but produced as an effect of the play of signifiers, according to the dual dimension of metonymy (concatenation, combination of one term with another) and metaphor (substitution of one term with another). In other words, meanings do not exist independently but are produced by the sliding of signifiers, however, as this play of signifiers is potentially open-ended and infinite, for signification to emerge there must be a quilting or halting operation.

This means that some signifiers are more prominent than others and can halt the endless sliding of signification. These anchoring points retroactively fix the signifying chain, and thus its semantic ambiguity, by giving it apparent stability and the illusion that reality is consistent (Fink, 1999; Klepec, 2016). Lacan calls this anchoring point the Master Signifier, represented by the matheme S1, which is what establishes and produces seemingly consistent significations: as quilting points, these stabilise and fix the signifying chain – albeit temporarily – while the signifying chain, indicated by the matheme S2 and representing 'knowledge', depends on this dominant signifier. As such the Master Signifier (S1) establishes power relations because it 'commands' the signifying chain and organises the way in which the political world operates (Kapoor, 2020, 6) by ultimately naturalising meanings. For example, it is the anchoring function of the Master Signifier in representation that naturalises the meaning of signifiers such as *climate change, mitigation, ambition, efficiency* through which we understand, establish, and maintain our societal, economic, and thus climate relations between stakeholders, political institutions, and citizens. The Master Signifier can be recognised in the text as that 'special word' that, as it were, makes things work, holding authority over the entire field of signification. For example, in the next chapter, I will point to how climate action mitigation knowledge (S2) – firstly presented in a spirit of 'free enquiry' – is in fact put to work in relation to a command (S1) of competitiveness, accumulation an ultimate growth, where the role of the 'consumer' cannot be questioned. Therefore, the relations between signifiers as overarching socio-linguistic structure determines the social link or social bond – which is for Lacan another name for 'discourse' – that keeps our intersubjective societal relations together.

Discourse: the binding structure

The previous introduction to Lacan's understanding of the process of signification helps us appreciate that the relationship between signifiers – the chain of signification – constitutes an overarching socio-linguistic structure that allows us to establish and organise our societal, economic, and ultimately climate relations, that is relations that make climate change knowable and governable.

According to Lacan, discourse is precisely the signifying structure that shapes, sustains, and forms our lives as 'speaking beings' (*parlêtres*) (Lacan, 2007, 13). This powerful linguistic structure affects and conditions the entire life of the speaking subject, in actions, affects, thoughts, customs, and daily praxis (Klepec, 2016, 115–118) and for this reason Lacan refers to discourse as 'social bond', or 'social link' (from the French *lien social*). The discourse can clearly exist without words, but it subsists in those fundamental relations that need language to be established and maintained (Lacan, 2007, 13): in this respect, the discourse has a formative function and there is no pre-discursive or extra-discursive reality because we cannot make sense outside language. Instead, every reality is defined discursively, and the subject has no choice but to resort to socio-linguistic networks of signifiers – what Lacan calls the 'Symbolic' – available to them to create sense and more widely to establish, maintain, and regulate any intersubjective societal relations. This includes how we, as speaking subjects, make climate change phenomena intelligible and governable, that is how socio-political actors respond to climate issues. In other words, we as speaking beings cannot escape the Symbolic order:

> It is from the fundamental relation (the relation from one signifier to another signifier) that the subject emerges, via the signifier which represents this subject to another signifier.
>
> (Lacan, 2007, 13)

The subject is thus structured in relation to a socio-symbolic network that constitutes for them what Lacan calls an 'Other', that is, a 'locus in which is situated the chain of the signifier that governs whatever may be made present of the subject' (Lacan et al., 1998, 203; Lacan, 1964). By way of example, the EU's climate change policy-making process and governance manifests itself as a multi-institution, multi-stakeholder, and multi-level process which in Lacanian terms constitutes an interpellation of the Other, the activation of a process of identification with the socio-symbolic order. This interpellation includes all the instances by which a subject presupposes knowledge – meant as what needs to be done to organise and maintain given climate policies – ranging from the micro practices of how EU officials work in their offices and interact with actors inside and outside the European headquarters to more visible outputs – such as a policy or legislative outputs and/ or any intra-institutional and inter-institutional negotiations and consensus seeking practices.

However, on the one hand, the real subject presupposes symbolisation to make sense and establish and maintain societal relations, on the other hand, the discourse presupposes a real speaking subject to be maintained – whether EU representatives, the stakeholders, the citizens, who are all bound together as representatives of this structure and all defined by this structure.

As we shall see in the second half of this chapter, Lacan, in fact, does not envisage one social bond, but four main social bonds, and a fifth variant. At the same time, it is impossible to appreciate the theory of the four discourses without addressing the novelty aspect of Lacan's theory, that is the relationship between discourse and subject, which is what sets him apart from other established poststructuralist discursive approaches.

The Lacanian subject: the subject of the enunciation

The peculiarity of the Lacanian subject can only be understood by considering the paradox of language: entering the medium of language, that is the act of speaking, always produces an ambiguity of signification, which is inerasable and defines us as humans. This ambiguity is perceived as loss (*perte*) in the enunciation, as the socio-symbolic order can never fully complete the speaking subject. In other words, for Lacan language comes from this fictitious yet necessary socio-symbolic network – the Other – but the inevitability of choosing a language, namely of identifying with language, generates an irreducible lack or cut within subjectivity, which originates from the priority of the (meaningless) signifier over the signified and the nature of the symbolic order (Lacan et al., 1998, 204–205; Lacan, 1964).[1] It is because of this ontological lack as residue of signification that the signified (meaning) slips metonymically across the signifiers and full signification is always deferred (Stavrakakis, 1999, 28–29).

Lacan explains this excess of meaning perceived as a loss with a small object, the *objet petit a*, that is, a conceptual tool that stands for the unattainable cause of desire, with its liberating potential (Lacan, 2007, 13–18), where 'liberation' refers to the possibility of a paradigm change. We can think of this object as the cause of a never-achieved state of wholeness, fullness, or plenitude: as Burgess puts it, it is not an object at all: 'it is the cause of desire but not the actual object of desire' (Burgess, 2017, 659). As Kapoor observes (2020, 36–37), desire in Lacan is not positively produced as in Foucault, but it is triggered by the lack that 'cuts' the subject when entering language, so it is an excess and loss to the discourse. More to the point, this small object, which stands for the lack that triggers desire, embodies the impossible 'full satisfaction' that is lost while enunciating (Stavrakakis, 1999, 49–53) and which results in a paradoxical (dis)satisfaction that Lacan calls *jouissance*. In other words, entering the symbolic realm of language – the linguistic system – entails the loss of *jouissance* or, rather, the achievement of *jouissance* as loss, and, ultimately, the endless deferral of the promise of full satisfaction. The key issue here is that this loss is fictitious – there was no real primordial full fulfilment: as a subject we just try to regain and recapture this fullness but end up in endless dissatisfaction, with the promise of enjoyment always deferred (Kapoor, 2020, 14–15).

Ultimately, the subject becomes the locus of an impossible and inconsistent identity (Stavrakakis, 1999, 13–17) constituted through its own lack: it is for this reason that Lacan speaks of a split, barred, alienated subject, represented by the matheme $ (Lacan, 2007, 13), or deploys the metaphor of 'castration' to emphasise how subjectivity is based on a failed identification and as such is always partially constituted and therefore incomplete (Newman, 2004, 145). In other words, the split refers to what is produced between the socio-symbolic structure and the ever-present excess that makes 'wholeness' impossible and it can be concluded that the Lacanian subject coincides with this inconsistent object, thus it is identical to their lack.

Indeed, Evans explains (sic):

The very word 'I' (Je) is ambiguous; as SHIFTER, it is both a signifier acting as subject of the statement, and an index which designates, but does not signify, the subject of the enunciation. This is how the subject is divided between these two levels, namely in the very act of articulating the I that gives the illusion of unity.

(Evans, 1996, 56)

In this ontological residue of signification, we see more clearly how Lacan integrated the structuralist linguistics of Ferdinand De Saussure and Roman Jakobson with Freud's psychoanalysis by providing a new, original theory of the unconscious. We resort to language to make and create sense, and to constitute ourselves, but every meaning brings with it a fundamental ambiguity, which is precisely where unconscious drives inscribe themselves:

The unconscious is constituted by the effects of speech on the subject, it is the dimension in which the subject is determined in the development of the effects of speech, consequently the unconscious is structured like a language.

(Lacan et al., 1998, 149)

In the above quote, we see that Lacan ascribes an 'ex-timate' (internal *and* external) linguistic dimension to the unconscious and reinterprets it in conjunction with the socio-linguistic system that constructs and determines the subject as a *parlêtre* (speaking being) and thus forms our social reality (Evans, 1996, 220; Tomšič and Zevnik, 2016, 2). This is because the unconscious for Lacan is not a hidden realm of unpredictable drives, but rather, a linguistic place in which desire manifests itself (Hook, 2013), it is sociocultural and conceived intersubjectively, it is part of our subjectivity but resides outside insofar as language precedes us and we form our subjectivity through the Other. The Lacanian subject is therefore the (split) subject of the enunciation, the subject of the unconscious produced by language as a surplus/lack of sense.

With this emphasis on the speaking being and its state of impossibility, it can be argued that this lack is what holds discourse and subject together, or that desire fixes the subject by anchoring it to a signifying chain that provides them with some (temporary) fixity and the subject becomes libidinally invested in it (Kapoor, 2020, 37). Consequently, Lacan' s ontology of the subject strikes as a middle ground between an essentialist universality and Foucault's politics of difference, with identities and power configured around an abstract symbolic place (the universal of the structure) – even though this is lacking – and a particular dimension which fixes political identities, yet only partially (Newman, 2004, 151). This makes it possible to conceptualise a different political subject as a split speaking being – whether the EU representatives, the stakeholders or the citizens – who, in the pursuit of given climate objectives, interpellate the Other and become caught and invested in a signifying machine determining the climate relations available to them – along with the excess of meaning thereby produced. Notably, it is precisely the question of how this lack-excess is mastered that leads us to reflect on the nature of the EU's climate transition and on what constitutes potential change. Indeed, from a Lacanian viewpoint, this meaningless remainder generated by symbolisation is the only access to change we have via disrupted signification and is rendered visible in gaps, weaknesses, and blind spots that 'crack' the discourse and make it look always partial and inconsistent. The alternative is, strictly speaking, conservative: lack is neutralised and integrated within the dominant type of signification, for example, by being turned into commodified knowledge and consumption objects.

Symbolic, Imaginary, Real: the three registers of the discourse

In sum, the subject holds together three registers of the discourse insofar as these are presupposed in enunciation. First, we have the already mentioned Symbolic, which is the socio-linguistic network in which we are embedded and to which we resort in our intersubjective relations. For example, if I observe or interview an actor on the EU's climate action, the signifying chain they deploy, which is made of 'knowledge', 'modelling', and 'targets', or 'efficiency', is the socio-symbolic network that forms our way of understanding and speaking of climate policy. The second register is the Imaginary, the realm of meaning, that is, our individual mental representations of a given signifier. Mental representations can be similar, but not identical for everyone (Pavon Cuellar et al., 2010, 2). For example, the signifier *renewables* can produce different mental representations: I can think of solar panels (the technology) or even a type of energy (solar energy), while my interviewee can think of wind turbines (the technology) or a different type of energy (wind energy). While I can know my own Imaginary, I cannot be sure of the Imaginary of the person we address. In any case, even though it is not possible to know the exact mental representation, meaning is reduced to sense, a content that is similar but not identical. For example, in relation to *renewables*, this would mean that we consider them as a 'set of energy sources that can be naturally replenished'. Finally, there is the Real, which is non-symbolisable and yet produced by the Symbolic (language) the moment the subject speaks and enters the medium of language. The Real is the realm of *objet petit a* and *jouissance*, that which escapes language representation in that it is produced by the discourse the moment the subject speaks, as a surplus, nonsensical remainder. An example of that can be detected whenever representation breaks down through gaps, weaknesses, and contradictions that make a seemingly consistent discourse fractured and inconsistent. Newman (2004) asserts that Lacan's Real has political implications: as it falls outside the Symbolic but is ultimately produced by it and sets in motion a series of (failed) political identifications, it also exerts hegemony because it generates the subject's desire to fill this lack. In other words, we are situated, and we constitute ourselves in all these three registers 'with the Symbolic and Imaginary helping to make the fabrics of our reality (however incompletely), and the Real tearing them apart' (Kapoor, 2020, 6).

Fantasy and jouissance

By stating that we constitute ourselves in the discourse does not mean that subjects all believe this discourse or are convinced by the discourse. Rather, the discourse provides a suturing function which the subject desires to believe to 'avoid the anxiety of the gap between the Symbolic and the Real' (Morrison, 2003, 278) – that is, the realm of the socio-linguistic net and the realm of *objet a* and *jouissance*, respectively. This suturing function pertains to what Lacan calls fantasy (*fantasme* in original) which is a construction located in the socio-symbolic order that promises to eliminate lack and to cover the impossibility of full representation by achieving a given desired object. This means that the ontological lack or void produces anxiety in the subject who then engages in an endless quest for the resolution of this anxiety, and this leads to unattainable narratives that apparently promise to resolve this lack. The key point is that its realisation is always deferred and lacking: every fantasy formation is articulated around the *objet petit a* (object qua lack) and therefore has implications in the Real, with the implication being *jouissance*. More to the point, the missed realisation of the fantasy is exactly what sustains the promise of fullness, it functions as a support for a desire to identify with (Fletcher and Rammelt, 2017; Stavrakakis, 1999, 46–51) and keeps subjectivities suspended and animated by temporary manifestations of enjoyment (Mandelbaum, 2022, 8). For instance, the concept of fantasy explains the myth of 'decoupling' – that is the idea of separating growth from environmental pressure – and its reassuring function (Fletcher and Rammelt, 2017) but has a broader application in the field of international politics. For example, Mandelbaum (2022) makes the case for the political dimension of fantasy and *jouissance* to explain the modern iteration of nationalism beyond the (Foucauldian-inspired) discursive practices that constitute the Self of the nation-state. More to the point, nationalism acts like a fantasy which provides a utopian and narrativised response of wholeness and security to the split subjectivity by identifying an outside, othering element that is to blame for their inability to achieve that (utopistic) fullness (Mandelbaum, 2022, 14–20). What animates the fantasy of nationalism as the 'lost thing' that needs to be retrieved and recaptured condemning to repetition and failure is however its affective dimension, namely *jouissance*, that temporary partial enjoyment which binds our body and minds together 'as a nation' (Mandelbaum, 2022, 15–22).

Discourse and subject have been regarded so far in general, ahistorical terms, as if there is only one possible social bond and one subject. In fact, according to Lacan, there are four possible social bonds. This claim corresponds to the historicisation of the discourse articulated by Lacan, especially in his seminar XVII

(1969–1970), which, together with the previous one, constitutes the most socially and politically explicit part of his entire oeuvre. The theory of the four discourses is therefore particularly relevant in this work for the ways in which these conceptual tools can be used as mind maps, insofar as each discourse describes a different type of knowledge underlying different power relationships, as well as a different subject.

The theory of the four discourses

The fixed framework and the four discourses

In seminar XVII (1969–1970), Lacan elaborated his theory of the four discourses, four conceptual apparatuses rendered through a graphical representation. Introducing the theory of the four discourses is useful because the four discourses are historically determined (Campbell, 2016, 241; Feldner and Vighi, 2015, 71), at least inscribed in the historicity of Western modernity (Žižek, 2006, 109) and underpinned by Lacan's thoughts on the knowledge of modern science and capitalism in the aftermath of the events of May 1968 in France. It is at this stage that the link between Lacan's theory of discourse and political analysis becomes more explicit.

Although Lacan did not explicitly address what made these given social bonds emerge and when, and whether other types of bonds might emerge in the future, he claimed that a change from one social bond to another could be caused by a disruption that, as he puts it 'hystericizes' the discourse (Lacan, 2007, 35). These four discourses can be used as theoretical guidelines to understand the EU's climate action, insofar as they explain more clearly the relationship between subject and discourse and what does not work in the discourse. In other words, they help make sense out of the 'knowledge objects' that are key to this climate transition, by means of detecting any rupture in the discourse and understanding how these are handled, which ultimately helps us to understand whether we are facing a change of the discourse.

These conceptual tools are represented through a framework of four fixed positions, where the upper part represents the conscious dimension while the lower part stands for the unconscious, the repressed or disavowed (see Figure 3.1). This framework can be regarded as a scheme of communication (Klepec, 2016): on the top left, there is the agent of the discourse – the message sender; on the top right, there is the other – the receiver of the message sent by the agent. Agent and receiving other are not necessarily a person but a locus, for example, any EU institution

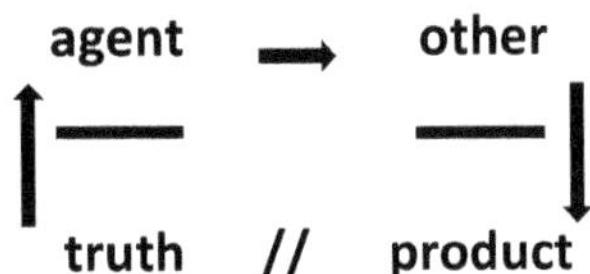

Figure 3.1 The fixed framework (based on Lacan, 2007)

such as the Commission or any UN locus such as the UNFCCC Secretariat. On the bottom left, there is the unconscious truth of the discourse, and on the bottom right, we find the product-loss of the discourse. The position of unconscious truth on the bottom left reveals that the agent is never fully in charge of the discourse.

The four discourses emerge through a quarter anticlockwise rotation in the fixed schema above. It is through this formal representation that we see the convergence of all the elements introduced so far. These are the Master Signifier qua anchoring point of representation, indicated by S1, which fixes and naturalises signifier and signified; the signifying chain as knowledge, represented by S2; the divided subject $; and the remainder (object cause of desire, i.e., *objet petit a*) as the excess-loss of the process of enunciation (Lacan, 2007, 32).

As shown in Figure 3.2, these discourses describe different types of social bonds and all leave a gap and contain an impasse, in other words, they are open and partial despite appearing closed and totalising:

> In supposing the formalization of the discourse and in granting oneself some rules within this formalization that are destined to put it to the test, we encounter an element of impossibility. This is what is at the base, the root, of an effect of structure.
>
> (Lacan, 2007, 45)

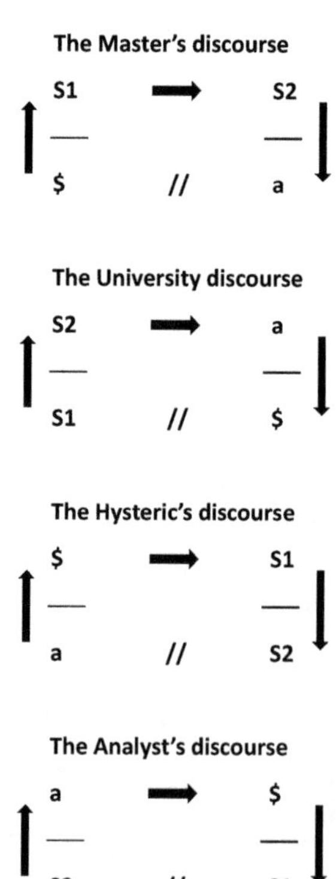

Figure 3.2 The four discourses (based on Lacan, 2007)

As each social bond describes different types of knowledge underlying different power relationships and different subjects, Bracher explains (1993) that we can understand the differences between Lacan's discourses by looking at the social effects they produce. The Master's discourse is the discourse of power and command; the University discourse is the discourse of knowledge, education, and indoctrination; the Hysteric's discourse is the discourse of desiring, questioning, and challenging; finally, the Analyst's discourse is the discourse of psychoanalysis, which is driven by the potential for transformation (Bracher, 1993, 53). The Master's and University discourses describe in different ways the reproduction of some form of domination. On the other hand, the Hysteric's and the Analyst's discourses are supposed to challenge the status quo, potentially generating real change by producing the desire for different significations. In the early 1970s, Lacan complemented this theory with a fifth discourse, the discourse of the Capitalist, which revolutionised the logic of the four discourses (Lacan, 1972, 32–40). This discourse constitutes a variation of the old Master's discourse and does not follow the standard anticlockwise quarter turn, but it is obtained from the Master's discourse by inverting the barred subject $ and the Master Signifier (see Figure 3.3).

For the purposes of this case study on the EU's climate change mitigation action, these theoretical guidelines exemplified by the four discourses can be considered in pairs, rather than in isolation. Indeed, throughout my analysis, I will place a special emphasis on the opposition between the University discourse of hegemonic knowledge and the Hysteric's discourse of 'real knowledge', and on the parallelism between the University discourse of dominant knowledge and the Capitalist discourse of commodified enjoyment (*jouissance*). Within this framework, the opposition between the University discourse and Hysteric's discourse allows us to reflect on how evidence-based knowledge in the EU's climate mitigation action is performed. Similarly, the parallelism between the University discourse and the Capitalist discourse allows us to reflect on how the impossibility of the discourse is turned into commodified knowledge and consumption objects of identification.

The University discourse: knowledge as command

The University discourse is a conceptual apparatus that describes a social bond in which knowledge (S2) is in command. Indeed, as we can see in Figure 3.2, knowledge takes the position of the agent of discourse and constitutes its driving force.

The University discourse indicates for Lacan the hegemony of modern science, thus a specific type of knowledge that asserts itself as neutral, measurable, quantifiable, bureaucratised, rational, and objective. However, the agent of the discourse is always commanded by an unconscious truth, which is the real and hidden engine of the discourse. In the case of the University discourse, this unconscious truth in the bottom left is occupied by the hidden Master Signifier S1, the signifier that naturalises the social bond. By following the logic of this discourse, we realise that this apparently neutral knowledge delivers, in fact, partial truths, in that the knowledge qua agent S2 is not as neutral as it seems, but there is an unconscious power relationship represented by the Master Signifier S1. In this respect, in Lacan's University discourse, a form of disavowal seems to be at play in relation to mastery.

Indeed, this discourse appears insidious in concealing the authoritarianism that is instead explicit in the Master's discourse through a command visibly imparted from the position of the agent. It is as if the University discourse disavows its performative dimension, 'presenting what effectively amounts to a political decision based on power as a simple insight into the factual state of things' (Žižek, 2004, 394). Hence, the University discourse conceals the relationship S2/S1, which is that of power and knowledge, and hides the performative and fictional character of knowledge under a flat and apparently objective knowledge, whereas in fact this knowledge works for the Master Signifier. As Pavon Cuellar et al. (2010) argue, this University discourse pertains to all socialist bureaucracies and liberal technocracies. By the same token, it is this presupposition that shapes our free-thinking, our analysis, our allegedly unbiased scientific inquiries but also our liberal ideologies and political orthodoxies under the influence of free market principles (Pavon Cuellar et al., 2010, 264–265). For example, the University discourse comes to mind when we think of the EU as a supranational organisation whose historical goal and 'core business' (Pelkmans, 2016) is the creation of the Single Market. Thus, it is this presupposition that shapes, for example, their climate targets, our climate modelling practices, and their climate governance structure.

In the University discourse, the allegedly neutral scientific knowledge (S2) attempts to control and tame *objet a*, the surplus of meaning perceived as loss that sets desire in motion, by integrating it into signification and turning 'lack' into a consumption object. Therefore, this object is integrated into signification and loses significantly its disturbing, traumatic, and therefore transformative potential (Feldner and Vighi, 2015, 93; Wright, 2016, 142). As Solomon puts it, 'objet *a* represents the desire of the other to absorb whatever knowledge the agent offers' (Solomon, 2015, 58). As indicated by Lacan in Seminar XVII, knowledge in this social bond goes along with *jouissance*, the ever-deferred full satisfaction (Lacan, 2007, 67): knowledge is the vehicle through which *jouissance* is produced, mastered, transmitted, and commodified (Wright, 2016, 138–142). Notably, *objet a* as excess of sense remains, the fact of being in the place of the other (top right position as shown in Figure 3.1, the fixed framework) does not make it symbolisable.

As a result, after we understood how knowledge qua agent of the discourse attempts to neutralise this lack, something is produced in the University discourse (bottom right of the graph): the split subject $, or rather different, deeply fraught subjectivities, defined by this new set of signifiers, who are excluded from relating and acting upon the Master Signifier S1. For example, the EU mobilises its knowledge apparatus in the pursuit of climate mitigation objectives as a form of ultimate full representation. From this desiring position of having an effective climate policy these split, fraught subjectivities, as workers, and consumers themselves set 'knowledge' to work with the signifiers available to them. These subjectivities are the EU bureaucrats who are themselves, as employees, caught in a chain of consensus-seeking practices, the stakeholders negotiating and lobbying their targets with the policy-makers, as well as the citizens as consumers placed at the heart of the energy transition to a low-carbon future. Yet, if the University discourse as a discourse of knowledge is nothing but a disavowed discourse of power, is there any 'real knowledge'?

Knowledge: University or Hysteric's discourse?

I defined the University discourse as the discourse of knowledge, with 'knowledge' being the modern science of increased mathematisation and technologisation. However, although Lacan first associated the University discourse with scientific formalisation, he later dissociated true scientific work from this discourse. In fact, due to the increased mathematisation and technologisation of science, authentic scientific enquiry was associated instead with the Hysteric's discourse (Lacan et al., 1990, 19). Hence, Lacan draws a distinction between a 'science' that does not cover contradictions and paradoxes of knowledge and a 'science' that works for the Master Signifier, which is that of the University discourse. In this respect, Fink (1999) explains the difference between the two 'sciences' as follows:

> It implies that the kind of knowledge involved in the University discourse amounts to mere rationalization. . . . We can imagine it, not as the kind of thought that tries to come to grips with the real, to maintain the difficulties posed by apparent logical and/or physical contradictions, but rather as a kind of encyclopaedic endeavour to exhaust a field. Working in the service of the Master Signifier, more or less any kind of argument will do, as long as it takes on the guise of reason and rationality.
>
> (Fink, 1999, 37)

In fact, the Hysteric's discourse is a discourse in which the split subject $ dominates in the agency position and calls into question the dominant knowledge and thus the dominant power relationships (the Master Signifier S1). Although the Hysteric's discourse is the name of one of the discourses, this does not mean that a 'hysteric' subject functions only within this Hysteric's discourse. For example, as an academic or a stakeholder, the hysteric can function within the University discourse, but what changes is their efficacy, since the effects and shortcomings are decided within that specific discourse. In fact, each discourse facilitates some aspects while hindering others (Fink, 1999, 30). Lacan himself explained that: 'The desire to know is not what leads to knowledge. What leads to knowledge is – allow me to justify this in the more or less long-term – is the Hysteric's discourse' (Lacan, 2007, 23). He contrasts the former 'commanded' type of science with a more authentic scientific enquiry that needs to think of complex systems to gain a comprehensive understanding of reality, with all the difficulties and contradictions that this might entail. In this regard, my investigation led to the search for any hysterical subjects in the field and for any attempt to overtly question the all too often assumed evidence-based policy on which the EU policy-making rests and is heralded.

Related to this desire to know – S2 in the position of truth, which is not the same knowledge as that of the University discourse – the psychoanalytical discourse conceives the analyst as the embodiment of the subject's lack (*objet a*) in the agent position. The analyst interrogates the subjects in their division and sets the patient to associate with the aim of producing a new Master Signifier (S1) (Fink, 1999, 37–38), thus a real change in signification and a changed subject. This is the reason why, if the Hysteric's discourse is the discourse of real and impossible knowledge, the Analyst's discourse is the discourse of transformation and real change, which I would identify as the ultimate, perhaps utopic, aim of my analysis. Consequently, if the Analyst's discourse is the discourse of change via the Hysteric's discourse of real and impossible knowledge, the University discourse can be seen as a discourse of reductionist knowledge. Furthermore, if the University discourse and the Hysteric's discourse are in a way antithetical, the University discourse and the Capitalist discourse are instead complementary, in that the latter addresses more explicitly the relationship between subject, discourse, and *jouissance.*

The Capitalist discourse: enjoyment unbound

As anticipated earlier in this chapter, Lacan complemented the University discourse with an underdeveloped fifth discourse, the Capitalist discourse (Lacan, 1972, 32–40), because he claimed that 'the important point is that on a certain day surplus *jouissance* became calculable, could be counted, totalized' (Lacan, 2007, 177).

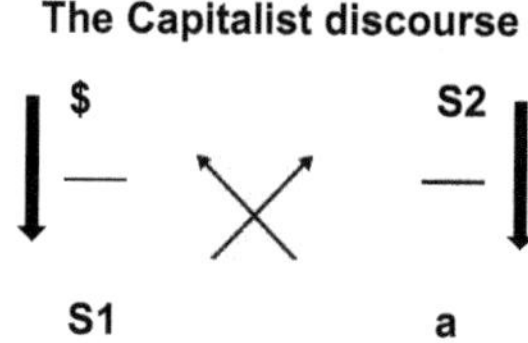

Figure 3.3 The Capitalist discourse (based on Lacan, 1973)

Lacan's critique is addressed to the fact that knowledge in modernity has been transformed into a countable and quantifiable entity, thus as the right hand of capitalism. Lacan's formalisation of this discourse occurred in the aftermath of May 1968, when he began associating capitalism with the University discourse. The students' protests in French universities ended with the adoption of the credit points system (*unitées de valuers*, units of value) and, on that occasion, Lacan warned against the fact that what looked like a revolution, or a subversion, was in truth a redistribution of tasks within the same system. What was operating in the transition from the old to the new master was in fact a transformation of the place of knowledge that led to its quantification, commodification, and rationalisation (Campbell, 2016, 141–142; Koren, 2014, 254; Lacan, 2007, 29–32; Tomsič, 2016, 158). This way, knowledge enters the circuit of commodities, and their agents – whether students, researchers, or teachers – produce and reproduce that system (Boni, 2014, 136). In other words, from a reflection on the university as the locus of knowledge production, Lacan expanded his social critique and turned to a more macro, systemic reflection on knowledge, its production and the rationality underpinning this production.

Having said that, in the Capitalist discourse, mastery, purported by allegedly neutral knowledge of quantification and valorisation, has become invisible but pervasive and more authoritarian than any other previous historical form of command. Lacan refers to it as a very clever discourse (Lacan, 1972, 47–48) as its greatest achievement is in fact the exploitation or industrialisation of desire (Lacan, 1973, 94). In the Capitalist discourse, the alienated and powerless worker-consumer subject $ is the type of subjectivity constituted by capitalist relations and occupies the position of the agent. These subjectivities address their lack by setting productive knowledge S2 in motion, articulated in the socio-symbolic Other. Thus, this subject is supported by the knowledge expressed by the signifiers available to them (S2) in the hope that their (endless) desire can be satisfied. On the other hand, the subject is characterised by a perpetual desire and demands objects of enjoyment that fuel this lack without ever satisfying them. However, as the directionality of this discourse shows, these split subjects $ address their lack through the (concealed) Master Signifier S1, which is the pervasive power of the market (S1). This means that they are in truth subjected to an unconscious command, the same command of the University discourse. Thus, rather than a fifth discourse, the Capitalist discourse is the replacement or transformation of the old previous relations of authority, power, and domination (Feldner and Vighi, 2015, 75; Koren, 2014, 254; Tomsič, 2016, 158).

The skill of the Capitalist discourse is that it exploits the structure of the desiring lacking subject as a means of endlessly reproducing itself. As Šumič (2016, 33) argues, the Capitalist discourse manipulates the subject's desire and reduces it to demand. This way this clever discourse creates the illusion that it can supply the subject with what the subject is lacking thanks to scientific development and the market.

Hence, if in the University discourse *jouissance* is accessed through knowledge production, in the Capitalist discourse lack is valorised and turned into a positive feature (Feldner and Vighi, 2015, 82–83; Tomsič, 2016, 158–160). If we look at the directionality indicated by the vectoral representation of this discourse (see Figure 3.3), the Capitalist discourse pictures a seemingly closed circuit of enjoyment for which there is no gap or excess, as if it intended to model consumers' satisfaction (Wright, 2016, 143–144). Thus, the system provides the subject with commodified objects of enjoyment, and this gives a temporary sense of plenitude that strengthens this position of enjoyment as a must, not as painful lack (Feldner and Vighi, 2015, 71–72; Šumič, 2016, 33). The ambiguity of this surplus-*jouissance* is well conveyed by the French *plus-de jouir*, where 'plus' means both excess and lack, and in the case of the Capitalist discourse it is equated with Marx's concept of surplus value (Lacan, 1972, 48). It is indeed this overlap between surplus value and surplus enjoyment (*plus-de jouir*) that enables the capitalist production of objects to capture and enslave the endless desire of the lacking subject (Šumič, 2016, 33). For instance, we will see surplus-*jouissance* in operation where I illustrate how a climate mitigation action that is confined to the signification space of 'energy transition' appears to be an illusionary change supported by the integration into signification of its traumatic points, such as the recent enthusiasm around the rebound effect – the fact that energy efficiency virtually equates with productivity as those saving are re-invested and do not lead to a decrease in energy volume – which was previously denied. Ultimately, the Capitalist discourse can be defined as the discourse of *jouissance*, where the impossibility of the discourse is concealed, but does not disappear. Therefore, when looking at the EU's climate mitigation discourse a key point is to look at how *jouissance* is manifested and 'trapped' and observe if there is any way in which it returns to be traumatically disruptive.

Climate change mitigation action: a Lacanian practice?

The enunciating act in climate action

The three interrelated registers of the discourse – Symbolic, Imaginary, and Real – as well as the four, or rather five, social bonds as theoretical guidelines help translate this conceptual apparatus into a research and analysis practice that is suitable for social and political research. My starting point is distinguishing between the statement and the enunciation. The statement – or enunciated fact – roughly represents language as an abstract system of linguistic signs – what arises from the socio-symbolic signifying structure to which the subject resorts and which is available to us. Conversely, the enunciation – or the enunciating act – represents what comes from the subject of the unconscious emerging from their unique discursive position with that 'excess' (Lacan, 2006), that is, the speech produced in concrete situations which carries the excess of meaning (see Pavón-Cuéllar, 2014). As a result, the enunciating act is the moment when the mutually presupposed relationship between discourse and subject – as embodied by the three realms of discourse – is manifested in the field.

The question 'who utters the sentence' must be considered carefully when we apply a Lacanian framework to empirical political research: the one who utters that sentence is not the subject itself but the subject insofar as they speak the discourse of the socio-symbolic fiction (the Other) to which they are bound. For example, an EU official is expected to utter a sentence in the name of the official policy line of the EU as well from a place, for instance, the EU Commission as an institution. Consistently, it is not important who we interview in terms of the hierarchy of the Commission or the Parliament. Lacan himself makes the example of the diplomat: when diplomats speak to each other they are purely representatives. They represent something whose signification is beyond their persona. In their conversations and interactions, they register what the other person conveys as pure signifiers, not what that person is (Lacan et al., 1998, 220; Lacan, 1964). Accordingly, despite their apparent diversity and their different role in the institutional pyramid or organisational charts, the EU representatives or stakeholders as Lacanian subjects are the speaking beings that, in the pursuit of given climate objectives, are caught in the same signifying machine determining the climate relations available to them along with the excess of meaning thereby produced.

Empirically, distinguishing between the enunciating act and the enunciated fact is useful then because it is in the gap between what the subject wants to say in the pursuit of climate objectives and what they actually say – the signifying chain mobilised – that we can reflect on the nature of climate action, detect any fractures, see how these are handled by the speaking subject, and ultimately reflect on what constitutes change. In this leftover, indeed, lies that 'lack' that sets desire in motion, which Lacan called *objet petit a*. This gap is rendered visible in the gaps, weaknesses, and blind spots that 'crack' the discourse and make it partial and inconsistent, as these are the breaking points of the discourse in which the process of identification with language, perpetrated by the subject, fails.

A Lacanian discourse analysis in action

When operationalising a given discourse theory into a discourse analysis, there are multiple ways of doing it. Foucauldian-inspired discursive approaches allowed researchers to identify meaning struggles and competing discourses shaping our socio-political life and how these perform certain truths in each time and inform policies (for instance Hajer and Versteeg, 2005, 177; Epstein, 2011), but not explain *how* and *why* a given discourse prevails (Leipold et al., 2019, 457; Solomon, 2015, 18–19). While a Lacanian discourse analysis perspective retains an understanding of power-knowledge relationships being discursively produced and perpetrated and might converge with similar approaches in detecting the hegemonic discourse of power, a Lacanian discourse analysis enables us to open up a seemingly locked discourse and expose the excess loss of signification (the Real) and its liberating potential precisely because of the role of the subject and because Lacan envisages a possibility of a transformative discourse alternative to the hegemonic discourse (the Analyst's discourse).

Starting from the 'enunciating act' in which the effects of language on the speaking subjects as well as their coping mechanisms can be detected, a Lacanian discourse analysis shows the *unintended* effects of language on the speaking subjects and explains how and if transformation and a change of paradigm can occur by pointing to and leveraging on the subject's unintentional coping mechanisms.

My version of LDA (see Tolis, 2023) builds on Parker's seven elements to analyse a 'text' from the perspective of the analyst consolidated in social psychology (2005, 2010; see also Hook, 2013) and retains some of these elements, such as the formal qualities of the text, the anchor of representation and the role of knowledge (2010, 167–172) as the visible patterns of the discourse. Yet, deploying a Lacanian discourse analysis with political actors requires some adjustments compared to a clinical setting insofar as the former implies overcoming the analyst versus analysand distinction, precisely because the speaking subjects observed and interviewed are not committed to an individual and personal therapeutic process. Moreover, the analyst versus analysand distinction is untenable as although both in-site observations and interviews help collect an extensive sample of enunciating acts about how climate policy is forged, these are two different methods. Conducting participant observations means observing concrete manifestations of the discourse beyond the researcher's control, and in which they accept to be involved as a passive, although not neutral, observer. Indeed, a stakeholder consultation itself, as a practice that fits into a policy-making process is an instance of the discourse that would occur regardless of the presence or absence of the researcher. By contrast, information retrieved through interviews is an ad hoc situation created by the researcher. In all these cases, as every reality is defined by the discourse and every subject emerges from the discourse, Lacan maintains that 'there is no meta-language' (Parker, 2010, 166) through which to explain language objectively which means that the researcher too is involved in the discourse and that it is not possible to stop the process of identification with the text in the very same attempt to make it meaningful. Thus, I can adopt the perspective of the 'Hysteric' by challenging the authority of the 'Master' when I question the signifying chain whenever I spot a contradiction during an interview; alternatively, I can follow the logic of the hegemonic discourse and observe the effects on the speaking subjects by seconding that chain of signification.

Building on the Lacanian informed assumption that the subject is speaking not their discourse, but the discourse of the Other, that is the socio-symbolic fiction, and in line with Lacan's prioritisation of the signifier, it is possible to map the unfolding signifying chain as knowledge in which our subjects are caught and which they put to work. For example, as a way of mapping the Symbolic in the case of the EU's climate change mitigation action, I traced the system of structuration of language such as equivalence and differentiation, words and phrases, grammatical associations and structures, the signifying sequence that binds together climate change, mitigation, knowledge, and energy efficiency. To analyse the role of knowledge and the subject's relation to this knowledge, I traced the points where knowledge emerges as a presupposition of sense which emerges when – for example – where an EU stakeholder event is welcomed as a rational knowledge exchange, animated by a spirit of free enquiry. Via this exercise, we are not trying to reveal any hidden meaning behind the surface of the text but rather explore how objects, subjects, and relationships are constituted in the surface of the text-speech (Parker, 2010, 158; Parker, 2005, 167).

As Parker notes (2005, 170), mapping helps isolate potential Master Signifiers, which is what anchors representation, naturalises, and gives cohesion to the discourse. The Master Signifier can be recognised as that signifier in function of which knowledge is put to work and a seemingly consistent discourse is produced. For example, in interrogating the type of knowledge that underpins the EU's climate action, I will contrast this alleged knowledge neutrality and free enquiry with other elements that contrast with this neutrality, such as the necessity to preserve the EU competitiveness, or the epistemological reflection on what constitutes evidence-based knowledge perpetrated by an academic and scientist at one of these events – the Postgrowth Conference. This way it is possible to establish if climate 'mitigation' is a dominant signifier that commands and sets the overall direction of the policy-making or alternatively if the knowledge it mobilises is in fact commanded by another authority.

Mapping the Symbolic and isolating potential Master Signifier(s) is also the starting point for disrupting a seemingly consistent signification and generate alternative understandings and different perspectives on the text. More specifically, it is precisely this exercise of disruption and disorganisation of the text that makes it possible to detect the surplus-leftover of meaning, rendered visible in gaps, weaknesses, and contradictions, in that the unconscious is what functions as lack, absence in the text (Parker, 2005, 169–171). As Kapoor notes, when it comes to spotting the unconscious, the emphasis here is on the surface as opposed to depth: we are not excavating unconscious desires from the back of individual minds, but we are detecting the 'unconscious hidden in plain view' (Kapoor, 2020, 14; also Hook, 2013). For example, *energy efficiency* as a presupposition of sense mobilises and keeps together a set of socio-political climate relationships between policy-makers, stakeholders, and citizens around delivering reductions in emissions. Yet, at different times of the enunciation *energy efficiency* is spoken of as a legislative act, as an equation, as a technology, as economic savings, and these are not always consistent with one another. This way the discourse of knowledge as represented by 'energy efficiency' results always as partial and its meaning deferred across the signifying chain, as an effect of the surplus-loss of signification produced in the enunciation. In short, a Lacanian discourse analysis emphasises the Real and the exercise of disruption helps detect the fractures where the discourse reveals itself as a discourse of impossibility.

However, when speaking of meaning, we must remember that it is impossible to know the individual single mental representation of the speaker, that is, their own Imaginary. Thus, we cannot search for imaginary similarities, but the signified (meaning) is reduced to the sense, meant as what the signifier and signified share, where they intersect (Neill, 2013, 339; Pavon Cuellar et al., 2010, 2–7). For example, when speaking of renewables, it is impossible to know if my interviewee is thinking of 'wind turbine' or 'solar panels' or the source of energy, but we share the fact of talking of a 'set of energy sources that can be naturally replenished and its related technologies'. Hence the main goal of an LDA is to generate possible alternative interpretations, to denaturalise the dominant representation and to open alternative possibilities based on the produced signifying patterns.

Importantly, this Lacanian-informed text displacement can be re-interpreted in conjunction with Lacan's theory of the four (or five) discourses. More to the point, opposing the University discourse to the Hysteric's discourse or establishing a parallel between the University discourse and the Capitalist discourse allows us to reflect on what underpins the EU's climate knowledge and how we understand the 'objects' of identification that are key to this transition, such as *energy efficiency, renewables,* or *circular economy.* Thus, it is by referring to the four discourses that we can observe how the emerging ruptures in the discourse are handled either by being positively integrated into signification or by disrupting it. Alternatively, we can explore if these fractures are at least able to shake the foundations of the discourse. This point can be exemplified by the case of *circular economy,* which was perceived as a disruptive element that would ideally point to reorganising our socio-economic relations and to the way in which knowledge is produced and exchanged: for this reason, it was strongly resisted, and the project was at first withdrawn. Interestingly, its reintroduction within the discourse, what I define as the 'acceptance' phase, has entailed a loss of the traumatic but liberating potential that 'circularity' carried with it, which might have produced a real change.

Conclusion

This chapter constitutes the theoretical backbone of this book and provides an in-depth discussion of Jacques Lacan's theory of discourse to set the ground for the study of the EU's climate change mitigation action in subsequent chapters. I explained how for Lacan signification proceeds through the interplay of signifiers that are retroactively fixed by an anchoring point, called the Master Signifier, which provides a relative but necessary meaning stability. It is this signifying chain that provides the subjects with an overarching socio-linguistic structure allowing them to establish and maintain all their intersubjective societal relations – that is what Lacan captures by reference to discourse as the 'social link'. The real speaking subject then presupposes symbolisation to create sense and more widely to establish and maintain these societal relations, an instance of which can be the multi-institution, multi-stakeholder, and multi-level processes that constitute the EU's climate policy-making. However, the activity of speaking always produces an excess of meaning which is perceived as a loss in the enunciation – Lacan explains this by reference to the *objet petit a* – which stands for the lack that stimulates desire, with its transformative potential. This object embodies the impossible, full satisfaction that is lost while enunciating and that results in a paradoxical (dis) satisfaction, called *jouissance*. It is because of this ontological lack – visible in the blind spots, contradictions, and weaknesses – that the signified (meaning) slips metonymically across signifiers and full signification is always deferred. This does not mean however that the subject is always consciously adhering to that discourse, but that they desire to believe it by resorting to socio-symbolic constructions, called fantasies, that promise to cover the impossibility of full representation, by achieving the fullness of *jouissance*. With a subject that ultimately coincides with their lack rather than with an essence of the individual psyche, Lacan's theory relevance for socio-political analysis emerges when we consider the socio-political production of subjectivity, insofar as this lack can only be filled by available social discursive representations conferring them a stable identity. This dialectical relationship between discourse and subject is further exemplified by the three interrelated registers of the discourse, the socio-symbolic network, the individual mental representation of the Imaginary and the Real as the realm of *objet a* and *jouissance*, which show how these are all held together by the Lacanian subject and emerge in the moment of the enunciation.

In this respect, Lacan's theory of the four discourses reveals its usefulness as these are intended as theoretical guidelines that allow us to reflect on what underpins knowledge in the EU's climate mitigation action, how we understand the 'objects' of identification that are relevant to this transition, such as *energy efficiency, renewables,* or *circular economy* and what constitutes real change based on how the lack they generate is mastered. Indeed, the four discourses can be differentiated according to the social effect they produce: the Master's (with its Capitalist variant) and the University discourses embody the dominant discourses of power and knowledge, while the Hysteric's and the Analyst's discourses embody the discourse of defiance, transformation, and ultimately change. As I argued, the opposition between the University discourse and the Hysteric's discourse constitutes a point of reflection as to what constitutes evidence-based knowledge in the EU's climate mitigation action. Instead, the University and Capitalist discourses are complementary in the way they deal with the impossibility of the discourse, namely by positively integrating it into signification and turning it into commodified knowledge and objects that are key to the current EU's climate action, in a vicious circle of (dis)satisfaction that we called *jouissance*. Finally, this theoretical framework can be translated into Lacanian discourse analysis, in which the starting point is the enunciating act, that is, the moment when the mutually presupposed relationship between discourse and subject is manifested. This is also the moment when all the elements of the discourse – that is the socio-symbolic and the excess-leftover of sense qua Lacanian subject – emerge.

A Lacanian discourse analysis consists of mapping the unfolding signifying chain as manifested in the enunciating act with the aim of tracing the presumed knowledge in which subjects are caught and which they put to work. On the one hand, this exercise – in conjunction with the conceptual guidelines provided by Lacan's theory of the four discourses – makes it possible to isolate potential Master Signifiers; on the other hand, and perhaps more importantly, it disrupts the anchored reading and exposes the Real, via the juxtaposition of how a given signifier or set of signifiers are spoken of at different times of the enunciation, by generating several different perspectives on the text. It is precisely this exercise of disruption and juxtaposition that allows us to detect the unsymbolisable excess-leftover of meaning that is visible in the gaps, weaknesses, and contradictions that make discourse partial and inconsistent. Ultimately, we can regard the ruptures in the discourse as the real ground for reflection on the transformative potential of the EU's climate action to the extent that we can identify how the traumatic points of the discourse are handled by the subjects, that is either by being positively integrated into signification as produced and transmitted knowledge and/or as consumption objects, or by defying and disrupting the hegemonic discourse, and ultimately producing alternative significations and thus real change.

In the following chapters, therefore, I will first disrupt a seemingly closed discourse of climate mitigation 'knowledge' (S2) with the aim of exposing the disavowal of the real authority of the social bond (S1). Then I will explore more in-depth the case studies of *energy efficiency* and *renewables* as derivatives of this knowledge at work, with the aim to explore their transformative potential via disrupted signification. Finally, I will narrate the story of a relatively new signifier which entered the scene of EU policy-making and initially disturbed the dominant discourse, that is *circular economy*.

Note

1 Žizek claims that the Symbolic opens up the wound it professes to heal (1993, 180).

4 Climate change mitigation action

Of knowledge and disavowal

Introduction

Starting from the logic of the signifier as detected empirically and in line with a Lacanian discourse analysis previously outlined, this chapter starts investigating how the EU's climate change mitigation action is thought and constituted by the speaking subjects. Within these spoken signifying structures, discursive interdependencies between climate, energy, environment, and the economy are constituted in relation to the period referred to as 'in-between strategies'. The aim of the chapter is to demonstrate that the evidence-based knowledge, on which the EU's climate change mitigation action is built, disavows in fact its real authority, its Master Signifier (S1). Despite this, a Lacanian discourse analysis shows how a seemingly locked discourse is, in fact, open, and thus makes it possible to disrupt the hegemonic representation to expose the fractures on which actors can potentially leverage to bring about real change. The chapter is structured as follows.

First, the EU's climate mitigation action is placed in the context of knowledge by its actors, which is hailed by the EU as evidence-based, neutral, and objective and is claimed as the driving force in climate mitigation action. At the same time, this factual assertion is held in tension with other events – such as an EU official public consultation – where this spirit of free enquiry is disturbed by language associations juxtaposing the fight against global warming with the state of the economy.

In expanding my Lacanian-based investigation to the everyday work of the EU representatives, I show that the subjects observed and interviewed are split between the powerful linguistic structures defining them and the desire to come up with an effective climate action. Importantly, these EU representatives emerge as facilitators of quantifiable and rationalised knowledge that is mutually reinforced by a rigid bureaucratic work organisation across policy areas beyond which is impossible to think and in which any alternative form of thinking that deviates from measurement and rationalisation is discarded. As a result, the nature of this allegedly evidence-based knowledge lies in the disavowal of its real authority, which is never questioned. It is this disavowed command (S1), embodied in the central role of the consumer in the transition to the low-carbon future, which constitutes the real authority of the signifying chain, that is the Master Signifier. This command, which is expressed for example by 'growing competitiveness', and as such points

DOI: 10.4324/9781003378228-4

to other chains of 'accumulation' and ultimately 'growth', constitutes – perhaps unsurprisingly – the logical boundary of signification, what gives apparent stability to signification within our social bond thereby creating the illusion that reality is perfectly intelligible.

Nevertheless, despite the apparent stability of signification space determined by climate change mitigation qua energy transition – as per the effect of the Master Signifier – we should not think of the discourse as a closed structure. To demonstrate that the discourse is in fact open and fractured due to the intervention of the subject of the (unconscious) enunciation, I disrupt the locked signification with the aim of generating possible alternative interpretations. First, I show that the sliding of *climate change*'s signification as a biophysical phenomenon into the energy signification space excludes in the speech all the other related human-induced environmental issues such as pollution, biodiversity loss, and ecosystem damage. Yet, these other interrelated environmental issues persist in fact as an excess-loss – at times defied by more progressive policy actors who warn about the disastrous environmental consequences such climate-driven energy solutions can have. Then, I continue by illustrating the partial and nonsensical character of *mitigation* where its meaning is perhaps dependent on its institutional character and the silos' signification spaces, which in turn are an expression of a bureaucratisation of knowledge and organisation of work. Finally, I provide an empirical account of how this allegedly evidence-based knowledge can be disturbed by scientists qua hysterical subjects, opening opportunities for a real change in a constraining social bond which resembles the social link described by Lacan in the University discourse.

Climate mitigation action: 'evidence-based' knowledge at work

My Lacanian discourse analysis starts with mapping the unfolding signifying chain as manifested in the enunciating act with the aim to trace the presumed knowledge in which subjects are caught and which they put to work. As previously discussed, Lacan affirmed the logical priority of the signifier over the signified (Evans, 1996, 189), based on which the signified is produced by the sliding of signifiers and signification emerges out of a quilting operation.

In the upcoming empirical exemplifications, the EU's climate action is presented by its speakers as a discourse of knowledge: in the speech, the knowledge (S2), informing an official consultation about the EU long-term strategy is welcomed as evidence-based, objective as well as legitimised by science and reason. Building on this, I expanded my investigation on DG Clima and DG Energy, the two lead DGs in the European Commission responsible for developing a climate change action. As we will see, the EU representatives emerged as facilitators of rationalised, quantifiable and valorised knowledge that is mutually reinforced by a bureaucratic organisation of work across policy areas – a 'silos' approach to work – and these set the direction for climate mitigation action. I conclude that the relationship between 'fact-based knowledge' – as expressed by quantification and rationalisation – and climate is more complex than it seems, and I open a line of enquiry that investigates the implications of this type of climate knowledge.

The official stakeholder consultation about the long-term strategy 'The EU vision for a clean, modern and competitive economy' (see Table A.2 in Appendix) was held at Université libre de Bruxelles (ULB). The event opened with a keynote by Alejandro Ulzurrun – at that time, DG Energy Communication and Interinstitutional Relations Head of Unit – who opened his speech with a link to 'university' as a metaphor for knowledge, as well as with a direct reference to 'university' as the place where evidence-based knowledge is produced and exchanged:

This is not a conference venue. *This is a university. And this is a choice. Debates have been held, fact-based discussions have been held, knowledge has been exchanged.*

(Alejandro Ulzurrun, 10 July 2018)

Pierre Gurdjian, the President of ULB goes further in sealing the social deal of climate action, university, and knowledge:

. . . dialogue illuminated by reason and science. The University has a key role to play in our societies to address *the problem we're all facing. . . . Universities are places where truth is upheld* as a core value. *Research infuses the rejuvenation of knowledge, where young minds are prepared for successful lives.* Places that are naturally linked with all facets of society. *Where ideas flow freely, inspired by free enquiry.*

(Pierre Gurdjian, 10 July 2018)

In both extracts, the unfolding signifying chain denoted a sort of legitimisation from 'science and reason' and 'free enquiry' that should inform the upcoming consultations with the stakeholders. As both the name of the event and the keynote speech from the (former) Climate and Energy Commissioner Miguel Arias Cañete suggest, this spirit of free enquiry and reason is immediately associated with the fight against global warming and that of the economy:

We are here to discuss the EU vision for an *economy that is clean and sustainable, more competitive and fit* for the twenty-first century. We are here to reach out to *all the stakeholders,* discuss *direction and speed of travelling to fight against global warming.*

(Miguel Arias Cañete, 10 July 2018)

Therefore, at this stage, we can keep our understanding open as to whether this consultation is about a climate-friendly economy or an economy-friendly climate. With this possible double interpretation of the knowledge that informs the EU's climate mitigation policy, I expanded on how the lead DG in climate action within the EU Commission, that is, DG Clima, works on an everyday basis with the help of interviews. On a general basis, I usually divided my interview into a procedural part and a policy-content part. The procedural part would set the tone for the unfolding metonymic signifying chain, in that the policy content answers were more likely to be affected by the role that specific individual covered. For example, a representative in DG Energy who deals with energy-related products and who talks about their role is unlikely to deploy a chain involving biodiversity, or pollution. The procedural part would set then the premise for looking at how the signifiers are deployed and, accordingly, I would tailor the subsequent policy-related questions based on the produced language associations.

The following extract from an interview with a DG Clima official starts from a procedural question, yet the language patterns produced their subjectivity based on the available objects of identification:

> My role in DG . . . is *supporting my hierarchy of policymaking in general with economic quantitative data* that typically involves working *with scientists at large that provide economic data* to see what they can basically send to us, what reports are there, and to translate that into policy relevant *analytical recommendations. We do that in this unit for what we call horizontal features. So, it's not a specific policy but it covers the whole economy.* That's why we try to do that. We do that both at EU level as well as the international level. *So, in my team a number of people look at EU's climate policy in aggregate: energy policy, agriculture policy, industrial policy, environment policy, employment policy, regional policy – cause it all matters in total if you want to address climate change. We try to the extent that is possible to quantify elements* of that and another part of my time follows international negotiations and *try to provide similar quantitative data there.* What other countries are doing? How does it relate to what the EU is doing? How does that action on a global scale add to or might be needed to achieve certain temperatures goals?
>
> (Interviewee 4, 12 October 2018)

As this DG Clima interviewee explained, the EU officials facilitate the exchange of knowledge that mainly consists of economic quantitative data so that they can be translated into relevant policy recommendations. From this extract, we have a better understanding of knowledge as a quantifiable and rationalised entity: 'we try to the extent that is possible to quantify elements' (Interviewee 4, 12 October 2018). The exchange is in turn mutually reinforced by a horizontal and vertical bureaucratic organisation of work across policy sectors: 'in my team a number of people look at the EU's climate policy in aggregate' (Interviewee 4, 12 October 2018). Similarly, the following extract, from DG Research and Innovation explains how the Commission work has evolved in terms of climate *aggregate policy*:

> I've been 20 years in the Commission. One of the evolutions is that *the work has become more and more horizontal, diagonal, coordinated, whatever word you want to use, but less and less in separated silos.* When I joined *20 years ago my first field of responsibility was called clean coal technologies for power generation. And that was it. I was not talking to transport, bioeconomy, or any other specific field.* So, things have changed a lot. . . . Not for the simpler as it has made *everything more complex,* but for the better.
>
> (Interviewee 9, 16 November 2018)

As we can appreciate from this interview extract, the evolution concerned overcoming as much as possible their strict bureaucratic 'silos' approach to work. However, we will realise throughout the book that this silos division unintendedly returns in the speech, and this suggests a possible reductionist conceptualisation of climate change itself. As the examples unfold, we notice how this quantified, rationalised, and bureaucratised knowledge apparatus is mobilised and expressed as a set of apparently neutral signifiers such as *quantitative data, fact-based, analytical recommendations*, and *aggregate policy*, that claim to be rational and objective and which set the overall direction to climate mitigation action. On the other hand, we might wonder whether these quantifications are indeed an expression of a true scientific enquiry. For this reason, the next two sections will further investigate the 'fact-based knowledge' that underpins the EU's climate mitigation action.

The disavowed dimension of knowledge

The extracts below show that the relationship between these quantifications qua knowledge and climate mitigation action is more complex than it might seem. In fact, it appears that thinking beyond this objective, quantified, and bureaucratised knowledge becomes impossible and that any alternative form of thinking that deviates from measurement and rationalisation is discarded. In this respect, in the following examples, I maintain that the nature of this climate evidence-based knowledge can be understood if we consider that a disavowal of a deeper injunction seems to be at play. More to the point, it seems that the EU does not want to impose any behaviours in the name of freedom and democracy, however, behind this apparent freedom lies a hidden command, regarded as natural and factual, that cannot be questioned. It is this disavowed order, embodied by the central role of the consumer in the transition to a low-carbon future, that commands in truth the signifying chain of climate change mitigation knowledge. More importantly, this emerges explicitly through an association between climate change mitigation as a low-carbon future and the energy transition, which has important implications insofar as it results in a reductionist conceptualisation of climate change mitigation itself.

In the following extract from DG Research and Innovation (known as DG RTD or R&I), the conversation shifted towards the relationship between knowledge and how to measure climate change mitigation:

So, to look beyond GDP, one problem is that it has several definitions. You can look at different things, *you can try and have another monitoring measurement or growth which would include GDP but also include externalities* of some of your activities like the cost of carbon, to take a famous example. And then you have a monitoring measurement of activity which is more than just growth domestic product that's one thing, *but can you extend it to many other things, way beyond GDP which do not have a monitoring measurement like well-being like are you happy? . . . But you're not going to tell me an answer in euro or in dollars. How do you measure this? How do you scale it? How do you compare it? How do you integrate it into one number or several numbers to measure growth? You could have GDP, you could have well-being, you can have education, happiness, there are so many ideas, some of which are purely in terms of monitoring units, some others are not that.* To agree on one is the level of difficulty, how do we compromise? *How do we do the trade-off between for instance GDP and happiness to take just two? How do we choose?* There's going to be a balance to find, and all of those difficulties probably will result in the fact the world is still *governed by GDP*, money, with all the distributional difficulties of it.

(Interviewee 9, 16 November 2018)

From this extract, we realise that the assumption of fact-based knowledge within the social bond is not as straightforward as we might think and it can be argued that thinking beyond this objective, quantified, and bureaucratised knowledge becomes impossible. This knowledge is so powerful that any alternative form of thinking that deviates from measurement and rationalisation is discarded: 'Are you happy? But you're not going to tell me the answer in dollars or euros, how do you measure this?' (Interviewee 9, 16 November 2018). Thus, the nature of evidence-based knowledge at work in climate policy is more insidious: it looks like it is factual and natural, but its real performative and political dimension is disavowed (see Žižek, 2004, 394). The disavowed dimension of knowledge emerged several times during interviews. For example, I asked a speaker from DG Clima whether the fact of imposing any collective change of behaviour, rather than a softer approach such as educating or raising awareness, would be more beneficial to climate objectives. The effect this question had on the subject is visible in the following extract:

The answer is no, probably it would be the contrary. *You probably would be seen as a dictator*, you want to dictate what other should do, should eat how fast they should move and *that is not going to work in a democratic society.* So, if you want to see more change in what people do in their own lives *in terms of their consumption patterns you need to educate you need to raise awareness.* But I think that, just *take as an example discussion about nutrition* and food. Think *of how much advice you would get good or bad of the 55 different ways of doing a diet. And you also have scientist who say it's all crap so* you think that an area where yeah people will have long discussions.

(Interviewee 10, 21 November 2018)

This quote demonstrates that as soon as I introduced an element of defiance or 'hystericization', such as a tougher approach for the sake of climate mitigation objectives, a degree of intolerance towards a 'command' emerged and the ghost of authoritarianism was invoked. This aspect leads to the question of whether a disavowed command is operative under apparent neutrality. In fact, on the one hand, it appears that the EU does not want to impose any behaviours in the name of freedom and democracy. Yet, on the other hand, behind this freedom and democracy mask lies a hidden and repressed command that goes unquestioned, 'their consumption patterns' (Interviewee 10, 21 November 2018), and is considered natural and factual. In this respect, what we observe here is the embodiment of Lacan's University discourse: knowledge (S2) is allegedly in command in the position of the agent, but it disavows its performative dimension, 'presenting what effectively amounts to a political decision based on power as a simple insight into the factual state of things' (Žižek, 2004, 394). This means that the power-knowledge relationship S2/S1 is concealed under a flat and apparently objective knowledge as well as its performative and fictional character, but in truth, this knowledge works for the Master Signifier, that 'special word' that makes things work, organises the way in which the world operates, and holds authority over the entire field of signification.

This disavowal of the performative dimension of the social bond seems to be constantly at play, as it emerges in the following extract from DG Energy when speaking of the 'real world':

> But the *constraints are real world constraints, they're not political constraints*, but it's not anybody. I mean, possibly, I would be able to persuade people to launch a campaign that says – wrap up warm this winter! But *I don't try to lodge a campaign* that says 'wrap up warm this winter' because that *doesn't make best use of what is distinctive about the Commission, which is law and money* and, 90 per cent law and 10 per cent money. And a law that says wrap up warm is the wrong thing you know, the future, trust.
>
> (Interviewee 8, 14 November 2018)

This speaker, besides restating the distinctive role of the Commission as a neutral, quantifiable, regulatory agent, quantified as '90 per cent law and 10 per cent money', led us to understand that we are in fact subject to orders, 'real world constraints' (Interviewee 8, 14 November 2018). Although we cannot investigate the individual mental representation of the speaker and see what their 'real world constraints versus political constraints' really are – the Imaginary register of the discourse – we can explore possible interpretations. For example, can the real-world constraints be the real degraded biophysical reality in which we all live? In this case, there would be no plausible reason for considering a degrading planet a constraint. Alternatively, can the 'real world' be the economic or financial constraints that regulate our lives in society? This seems to be a plausible interpretation that links back to what the DG RTD interviewee claimed when they spoke of a 'world is still governed by GDP' (Interviewee 9, 16 November 2018). It looks therefore that the social bond does not tolerate an explicit signifier in the commanding position (obey!) but this command must be shared in the democratic way of consumption, and this is associated with Lacan's University discourse. As Pavon Cuellar et al. (2010, 264–265) argued, this disavowed S2/S1 relationship is the presupposition that shapes our free-thinking, our analysis, our allegedly unbiased scientific inquiries but also our liberal ideologies, and political orthodoxies under the influence of free market principles.

More precisely, while in the Master's discourse, the relationship between knowledge and the Master is explicit because the Master Signifier S1 is in the position of agent and gives direction to the overall discourse, the University discourse is a Master's discourse in denial. This means that a Master Signifier (S1) commands the discourse of knowledge as a presupposed truth because the University discourse does not tolerate one signifier to monopolise the commanding position; rather, this disavowed position must be shared in a democratic way by the wisdom of signifiers on equal footing, under the façade of tolerance, freedom, objectivity, and political correctness (Pavon Cuellar et al., 2010, 264–265). As a result, if the Commission orders 'wrap up warm in winter' (Interviewee 8, 14 November 2018) for the sake of climate objectives, this would mean speaking from a position of the explicit Master demanding obedience, which is at odds with the ideals of freedom and democracy. In truth, the disavowed authority dictating consumption patterns exists and commands all the signifying chains (S2/S1). A similar example is provided by DG Clima, where instead of 'real world', a reference is made to 'real life':

> But then the next question is ok *what type of climate one wants to stabilize? How much effort we have to do now?* That's also quite easy 2 degrees – for a long time – and then, after Paris, we have to pursue effort –1.5 – that increases the need of ambition. So that's what we are working on and stakeholders they have different opinions. I mean if you go to sort of NGOs who are very, who only *base their view on equity* they ask *for, I would say, emissions reductions that are not achievable in real life.*
>
> (Interviewee 4, 12 October 2018)

In the above quote from DG Clima, something very similar to the 'real-world' constraints emerged and is rendered by 'emissions reductions that are not achievable in real life' (Interviewee 4, 12 October 2018). Again, the performative and political character of the 'world' manifested itself against its factual character. The 'real life' is perhaps not meant as the environmentally degrading planet that we live in, insofar as this cannot be regarded as a constraint to emissions reductions; but perhaps it can refer to the economic or financial constraints that regulate our social bond. This results in the paradox of performatively creating two 'fictitious' types of planets to stabilise: 'the next question is ok what type of climate one wants to stabilise' (Interviewee 4, 12 October 2018), as, in truth, we only have one planet. The effect of the disavowed command emerges more explicitly at Session I of the stakeholder consultation 'the EU vision for a clean, modern and competitive economy'. This session was called 'Cost -efficient ways for achieving a post-carbon European economy' and DG Energy's Director General, Dominique Ristori, explained the role of the consumer in the transition to a low-carbon future:

> . . . the world economic strategy as a real economic opportunity. This is clear when we're thinking of the profitability of all this exercise for the world population. And we're placing *the consumer to be put at the centre of the scene, the citizen at the centre of the scene. This will be part of a new economy of a new decentralized energy system.*
>
> (Dominique Ristori, 10 July 2018)

This emphasis on the role of the consumer is restated more strongly by DG Energy's Director Renewables, Research and Innovation, Energy Efficiency, Mechthild Wörsdörfer, at the same event during a session called 'Benefits of a low-carbon world for all Europeans – a Citizen's perspective':

> *This session is about consumers, all of us being consumers. How can we contribute to this clean energy transition, to the low-carbon technology decarbonization for the future?* So, we have an excellent panel here with a lot of different groups represented and we have also in our Clean energy package (2030 Ed.) a couple of *measures for consumers to make them more proactive, be it on energy efficiency, which is an obvious candidate for energy consumers, energy products, labelling where we can do more. It's also our Renewables Directive, which was voted this morning, where we have an article about self-consumption.* But we have many other examples in our Clean energy package, where we want to be *more proactive consumers and participants.*
>
> (Mechthild Wörsdörfer, 10 July 2018)

It can be argued that the untouchable role of the subject-consumer is emphasised as a core element guiding the transition to a low-carbon future. Furthermore, we can now observe more clearly a close association between climate change mitigation, as a low-carbon or decarbonised future, and the energy transition: 'This clean energy transition' (Mechthild Wörsdörfer, 10 July 2018). For this reason, the next section will briefly illustrate how the (necessary) anchor of representation gives relative meaning and stability to the discourse and, building on that, I will investigate and expose the possible fractures of the climate mitigation discourse as an energy transition.

The discourse's Master Signifier in plain view

Climate change mitigation action has so far been spoken of as a discourse of allegedly neutral knowledge (S2). This means that knowledge is in fact commanded by an order – the real and disavowed authority of the discourse (S1) – which gives the direction to the discourse and thus regulates our intersubjective relations in the climate field specifically, and in society more generally. This disavowed command has begun to emerge with a preference for the factual role granted to the 'consumer' over a tougher approach for the sake of climate objectives.

I will now turn to the command of the discourse and its role, what Lacan called the Master Signifier, to emphasise how it temporarily locks up the discourse and fixes its semantic ambiguity. More to the point, a Lacanian perspective suggests that this anchoring point of representation is necessary because it holds the whole discourse together, represents the logical boundary of signification and gives apparent meaning stability to the signifying chain and the illusion that reality is perfectly intelligible. In other words, without the Master Signifier, the social link would not function and would lose its symbolic efficiency. In the examples below, this authority, which can be expressed for example in the emphasis on a 'growing competitiveness', points to another signification chain of accumulation and ultimately economic growth. This is itself unsurprising and in line with previous discursive approaches that shed a light on the power-knowledge relationship, such as Hajer's (1995) early analysis of how the ecological modernisation discourse prevailed over the radical restructuring discourse. However, because of this anchoring point, the juxtaposition of climate action with energy transition becomes relevant, insofar as *climate change* and *mitigation* acquire meaning within this signification space, and this in turn results in a reductionist conceptualisation of climate change as a biophysical reality separated from the rest of the 'environment'.

In the previous extracts, the Master Signifier qua anchoring point of representation was at times hidden and was not mentioned in the 'text', however, it can still be recognised in the text as that something which makes things work. For example, DG Research and Innovation (known as DG RTD or R&I) and DG Clima published a 160-page report called 'Final Report of the High-level Panel of the European Decarbonisation Pathways Initiative' (EU Commission, 2018), shortly after the release of the 2050 long-term strategy. The high-level panel is composed of experts and institutions outside the Commission and in their Executive summary we read that the DG RTD Commissioner Moeda requested the High-level panel members to work on the following question:

> What strategy to adopt in Research and Innovation *in order to speed up and foster mitigation policies* in the EU that respond to the goals of the Paris Agreement, *while growing competitiveness of the EU Economy?*
>
> (High Level Panel report, 2018, 18)

As we can see the question points to the 'growing competitiveness' as the logical limit of signification, which logically excludes any policies and actions that would not increase the EU competitiveness. In turn, competitiveness would point to another battery of signifiers of production, consumption, and accumulation, and ultimately, economic growth, as the anchoring point of our social order. In fact, in the long-term strategy Communication (COM (2018) 773 final) the halting, anchoring function of modernisation and competitiveness is made explicit:

> The European Council, in June 2017, strongly reaffirmed the commitment of the EU and its Member States to swiftly and fully implement the Paris Agreement, underlining that the Agreement *is a key element for the modernisation of the European industry and economy. . . . The EU, responsible for 10% of global greenhouse gas emissions, is a global leader in the transition towards a net-zero-greenhouse gas emissions economy.* Already in 2009, the EU set its objective to reduce emissions by 80–95% in 2050. *Europeans have managed to successfully decouple greenhouse gas emissions from economic growth in Europe for the past decades.* Following the peak in EU greenhouse gas emissions in 1979, *energy efficiency, fuel switch policies and the penetration of renewables reduced emissions significantly.* In consequence, between 1990 and 2016, energy use was reduced by almost 2%, greenhouse

gas emissions by 22% while GDP grew by 54%. *The clean energy transition has spurred the modernisation of the European economy*, driven sustainable economic growth, and brought strong societal and environmental benefits for European citizens.

(COM (2018) 773 final, 4–5)

This extract is relevant for three important reasons. First, and on a surface level, it confirms the role of competitiveness, and its associated chain of consumption production and accumulation, as anchoring function of the social bond: 'Europeans have managed to successfully decouple greenhouse gas emissions from economic growth in Europe (COM (2018) 773 final, 4–5)'. More importantly, this anchor holds authority over the entire signifying chain and confers apparent meaning stability through the signifiers available to us and gives us the necessary illusion of the 'real life' Interviewee 4 mentioned in their interview with all the related discursive practices appearing as factual: 'emissions reductions that are not achievable in real life (Interviewee 4, 12 October 2018)'. Therefore, this Master Signifier is the actual anchor of representation of reality which represents the logical boundaries of the discourse; the 'something' that holds the social link – real life – together. Finally, and building on this point, we clearly see the semantic juxtaposition of climate action and energy transition that had already been mentioned in the case of the consumer at the centre of the energy transition in a low-carbon future. This is important because we can question how signifiers such as *climate change* and *mitigation* acquire meaning within this signification space and how *energy efficiency and renewables* are spoken in this signification space as concrete manifestations of this climate knowledge at work.

To summarise, the EU's climate action qua energy transition metonymy should not thus happen per se, but in a function of the modernisation and competitiveness of the EU economy and industry which is the discourse's Master Signifier. This boundary of the discourse has important implications because it suggests that *energy efficiency and renewables* – which will be the object of Lacanian discourse analysis in the next Chapter – could be concrete manifestations of this climate knowledge at work. Consequently, within these signification constraints, how can any subjective dimension emerge? In my previous theoretical exposition of Lacanian theory, I explained how for Lacan the subject is never completely over-determined by the symbolic structure and that the speaking activity always produces an excess perceived like a loss in the enunciation. This excess-loss to the discourse is the Real of the discourse and is represented by *objet petit a*, as the unattainable cause of desire, with its liberating potential (Lacan, 2007, 13–18) which can be approximately thought of as the cause of a never-achieved state of wholeness, fullness, or plenitude but not as the actual object of desire (Stavrakakis, 1999; Burgess, 2017, 659).

As a result, while the previous discussion was preoccupied with unveiling the disavowed Master Signifier of the EU's discourse of climate knowledge, I will now show how an apparently locked discourse is indeed open and fractured by virtue of the Lacanian (split) subject of the enunciation.

**Disrupting *climate change* and *mitigation*: the nonsensical character
of the signifier**

The signification space of climate change

The discussion conducted so far has placed the EU's climate mitigation action
within allegedly neutral knowledge. Within the discursive boundaries established by the Master Signifier of competitiveness, accumulation, and ultimately
growth – whose force is however disavowed – what apparently looks like neutral,
objective, and rational knowledge is in fact put to work in relation to an injunction
to accumulate, consume, and grow. At the same time, Lacan maintains a dialectical relationship between discourse and subject and therefore despite the apparent
stability in representation and closure of the discourse, the perfect intelligibility
of reality is an illusion. Instead, the discourse is open and inconsistent due to the
intervention of the subject of the enunciation – the subject produced by language as
a surplus of sense. As I will demonstrate, the subjective emerges when the speaking
subjects observed and interviewed are split between the powerful linguistic structures defining them and the desire to come up with an effective climate mitigation
response through the same shared socio-linguistic structure.

In this section, I argue that climate change mitigation takes shape as a
socio-political reality in the narrower signification space of climate + energy. This
climate + energy nexus in turn shapes and is shaped by practices such as the composition of the EU Commission, where DG Clima and DG Energy have the lead
role in climate policy. Moreover, it shapes and can be shaped by the actual policy
outcomes and their implementation such as the 2030 'Clean energy package'. Consequently, this closure of signification regarding *climate change* as a biophysical
phenomenon into the energy signification makes its signification always partial
because it excludes all the other related human-induced environmental issues such
as pollution, biodiversity loss, and ecosystem damage. This residue of signification,
however, is questioned by the perhaps more assertive stance of DG Environment
that warns about those political climate solutions that can have other disastrous
environmental consequences.

In the following extract, an EU representative from DG Research and Innovation speaks of an actual system change:

We try to find research which is of a more cross sectoral nature than before.
Research looking at not new specific technical devices but at socio economic
aspects, behavioural aspects, *and all the things that are required for a system
change* and that does reflect even the policy itself. Because when we wanted to
go from 5 per cent renewables to 20 per cent renewables *was still the same system with a few solar panels and windmills*, but *you plug your kettle in the same
old plug. Now we know that we run for a complete systemic change, a new
society a new system a new behaviour, new social relations because of* climate
change mitigation. We have to look at all of those aspects when we fund and
do research and not just take down the costs of the latest windmill generation.

(Interviewee 9, 16 November 2018)

The interviewed subject speaks of a 'system change' and a 'new society' that is not merely interested in adding a few more 'solar panels' (Interviewee 9, 16 November 2018). However, how is this desire for an effective climate action and system change spoken through the shared socio-symbolic structure?

To explore this further, let us start with *climate change* and *global warming* as signifiers. Climate change refers to a biophysical problem, the excessive temperature rise of our planet due to anthropogenic GHG emissions. For this reason, it is not surprising that climate change policy takes place within a signification space that is confined to energy. As an EU official at DG Environment explains:

> Historically the EU Commission climate policy *has been focused on energy.* This has some logic in that *75 per cent emissions come from energy.*
>
> (Interviewee 14, 29 November 2018)

This 'climate' and 'energy' deal is reflected in the composition of the EU Commission where DG Clima and DG Energy have the lead role in making climate policy and set the overall direction of policy-making.[1] We also see this work structure reflected in the stakeholder events and side events at the COPs around climate action, which are either hosted by DG Clima or DG energy (see Tables A.1, A.2, and A.3 in Appendix). Finally, the 'climate' and 'energy' nexus is reflected in the policy outcomes such as the 2030 strategy, as indicated in the title 'Clean energy package', as well as in all the references made to climate action qua energy transition. This is even more interesting if we consider that DG Clima is a relatively recent policy DG created from a branch of DG Environment – founded instead in 1981 – under the José Manuel Barroso's second Commission mandate (2010–2014). In the extract below, a DG Environment speaker explains that the policy-making activities of DG Clima are more in line with those of DG Energy than those of DG Environment:

> *We are not the lead DG on climate*, but our role is to check and discuss with colleagues some of *these political solutions that can have environmental consequences. . . .* For example, the trade off with biomass. Under the last few years *under the Renewable Energy Directive* that is in force now, *European countries have funded deforestation in third countries. DG Energy will edit this* in the new version of the Directive.
>
> (Interviewee 14, 29 November 2018)

As a result, climate policy seems to be signified in terms of energy policy and this 'closure' of signification has important implications at the level of the discourse as being spoken of by the real subjects. If we think of climate change as a biophysical phenomenon, this fits within a wider framework of environmental degradation widely speaking. For this reason, it should not be considered in isolation but in relation to other human-induced environmental issues such as pollution, biodiversity loss, and ecosystem damage, that is, the planet or the biosphere in its wider understanding. In fact, all environmental aspects are inter-linked, as a DG Environment official suggests:

Yes, it's all linked, *problems with biodiversity loss become more serious because of climate change and vice versa.*

(Interviewee 14, 29 November 2018)

DG Environment, which is responsible for these other environmental issues, acts in a way that tries to keep signification open to all the other environmental aspects. Although DG Environment claims to have a marginal role in climate policy, it acts as the watchdog that warns about those political solutions that can have other disastrous environmental consequences. Consequently, climate change mitigation takes shape as a socio-political reality in the narrower signification space of climate + energy that keeps its signification always partial, with the rest of the inter-related environmental issues as an excess-loss 'defied' as much as possible by the wider and perhaps more progressive stance of DG Environment. Therefore, within this signification space, how does *mitigation* acquire meaning?

Mitigation as nonsensical

The alleged meaning stability of the signifier *mitigation* can be challenged by exposing its ultimate 'nonsensical character' through the real subject of the enunciation and by exploring any emerging ruptures. 'Nonsensical' in this context does not mean that there are no GHG emissions to cut down, or no catastrophic global warming to prevent or mitigate. Rather, it refers to the way in which it is signified in the enunciation and the way in which it reveals each time its partial, never-achieved character that we can detect in the inconsistencies that embody the residue of signification between the enunciated fact and the enunciating act. We will see that the meaning of *climate mitigation* within the EU is perhaps dependent on its institutional character and the silos' signification spaces, an expression of a bureaucratisation of knowledge and organisation of work.

In the following extract, I mentioned the signifier *mitigation* to the interviewee from DG Energy working on renewables and the Renewable Energy Directive:

Because this is not so much mitigation, this is really achieving the target for a renewable energy.

(Interviewee 6, 7 November 2018)

From this assertion, we perceive a distinction between *mitigation* and *renewables* targets. As a result, I asked a question to clarify whether the two are related and in what way they are related:

Yes completely. But you know there are some Member States that said – oh well but we don't need renewable targets *we only want CO_2 targets* this is the only thing that matters. This is the position of some Member States, but this is not the position of the Commission. This is not perhaps the Renewables Directive. We say no, *renewable energy in addition to climate change has other benefits.* This is why there is a target for emissions reductions but there's also a target for renewables.

(Interviewee 6, 7 November 2018)

The discussion with the interviewee shifted here to 'targets'. This can be linked back to the pattern of quantification and rationalisation that characterises climate action's knowledge according to which each concept is constituted in terms of measurements, in line with Lacan's critique of how knowledge in modernity has been transformed into a countable and quantifiable entity, as the right-hand of capitalism. By following the interviewee's reasoning, *emissions reductions* and *renewables* are two different things as they are formalised by two different targets – as exemplified in the 2030 Clean energy package – with the interviewee associating *mitigation* with the first target (emissions reductions) but 'not so much' (Interviewee 6, 7 November 2018) with the other target (renewables). Therefore, prompted by the interesting associations, I asked the interviewee to clarify what *mitigation* in their view is (sic):

> *This is probably just institutional. . . . When I think about climate change mitigation, I think of the activities of DG Clima, because they do these things about mitigation,* about building dams, whatever is needed to cope with sea level rising or reforestation. This kind of stuff. And climate change preparedness, *all of this is for me linked to Clima. . . . Of course whatever we say the key argument is emissions reduction, but I look at it from the energy point of view. And for us this is renewables and maybe at some point CCS.* But, so for me, *emissions reduction is, this is for us the subpart of mitigation, which is what we are responsible* for, this is how I use the term but I'm not sure about anyone.
>
> (Interviewee 6, 7 November 2018)

A few issues emerged in these extracts from this speaker. This speaker had at first distinguished *mitigation*, perhaps meant as emission reductions, from *renewables*: 'but this is not so much about mitigation' (Interviewee 6, 7 November 2018). Thus, in the interviewee's mind, it is probably a matter of 'targets', and emissions reductions and renewables are translated into two different targets. This distinction is understandable as if we buy an electric car rather than a traditional diesel one, we are emitting less GHG emissions as the car is electric. However, if the electricity being used by the electric car is generated from either fossil fuels or renewables, this impacts the overall emissions reduction effects, which justifies the need for DG Energy to differentiate between the two different targets. However, if the renewables target is 'not so much mitigation' (Interviewee 6, 7 November 2018), the meaning of *mitigation* becomes dependent on its institutional character, 'all the activities linked to DG Clima' (Interviewee 6, 7 November 2018). Thus, the 'silos' approach unintendedly re-emerges in the speech despite being to some extent overcome at the conscious level, as claimed by Interviewee 9 'the work has become more and more horizontal, diagonal, coordinated, . . . but less and less in separated silos' (Interviewee 9, 16 November 2018).

Second, and perhaps more interestingly, the initial distinction between 'not so much mitigation' (Interviewee 6, 7 November 2018) is rendered more inconsistent as the speaker contradicts themselves by saying that GHG reductions – thus their mitigation – from that specific DG Energy unit's point of view is Renewables and the CCS technology, and that becomes at the end of the sentence a 'sub part of mitigation' (Interviewee 6, 7 November 2018). Whereas at the beginning of the conversation *renewables* were 'not so much mitigation', by the end these became 'a sub-part of mitigation' (Interviewee 6, 7 November 2018). At the same time, when referred to a colleague next door who accepted to answer a couple of questions as an energy efficiency expert, the same question is answered in a different way and *mitigation* is articulated as 'all measures that result in less emissions compared to a starting point. Not reductions in absolute terms' (Interviewee 6a, 7 November 2018). This shows how all the speaking subjects deal with the fractures in the discourse, the cut in subjectivity opened by the socio-symbolic network.

The discussion about how *mitigation* is spoken can be expanded even further. For example, another interviewee in DG Energy associated *mitigation* with *energy efficiency*, or rather energy consumption reduction, which logically originates from the metonymy climate action as energy transition in the signifying chain:

> Ok I'm . . . in energy efficiency and my job is to reduce Europe's *energy consumption*.
>
> (Interviewee 8, 14 November 2018)

In this conversation with one of the environmental stakeholders, namely CAN Europe, *mitigation* becomes something other than efficiency (targets) and renewable (targets), but remains further unspecified:

> What we have in common with these stakeholders is that we all believe that the EU should be more ambitious should *speak of higher mitigation targets but also energy efficiency targets, renewable targets etc.*
>
> (Interviewee 1, 19 July 2018)

Mitigation can also be controversial: in the LULUCF sector, *mitigation* acquires a different meaning in that emissions are not merely anthropogenic, but they are part of the natural cycle, so whether you count them or signify them as *mitigation* is debatable. Moreover, in understanding climate action in terms of targets and modelling it is difficult to tell what type of emissions are included in the modelling, as a researcher from DG JRC – a non-policy DG which is the scientific body of the Commission – notes:

> Because *whereas in the other sectors there are only emissions, it's all clearly anthropogenic. If you reduce emissions, you can count that as mitigation.* In the forest sector everything is much more complicated because CO_2 is absorbed by forests. *It's part of a natural cycle* that is perturbated by human action. It's very difficult how much of the emissions, how much of absorptions is due to human action or not. *The concept of mitigation in the LULUCF sector is much more complicated than in the other sectors, it's very controversial.*
>
> (Interviewee 15, 30 November 2018)[2]

We can expand that even further to see the associations that come up when thinking of *mitigation* or *mitigate*. In another interview with DG Clima, the need for mitigation is not, for example, a matter of planetary survival, but a cost-efficient action to undertake:

> the kind of narrative which we have been using at least for the last 25 years . . . that *we need to mitigate because that is going to be much cheaper than adaptation.*
>
> (Interviewee 10, 21 November 2018)

In the extract below from DG Grow it seems that mitigation spurs images of competitiveness for the industry. More to the point, *mitigation* is performed in function of the Master Signifier more explicitly:

> *Standards are primarily climate driven.* So, you have the *Paris agreement, you have what the EU would do to try to do and meet those targets* and the whole range of measures. But one of course will be reducing the amount of pollution that comes from transport which is one of the biggest contributors to pollution as well as GHGs and CO_2. So, you start from that that's all very well and good. *But what is the impact on, what is important to industry in Europe in terms of employment innovation? So, a subsidiary priority for these standards is to say – well this is happening all over the world.* China is particularly aggressive in pursuing. There's clearly a demand anyway for cars to be more, to be cleaner, *so one of the things by introducing standards is that you encourage industry to invest more in innovation in this area so that they can remain competitive* not only in the EU and meet the EU standards.
>
> (Interviewee 11, 22 November 2018)

Similarly, following competitiveness, *mitigation* is worth considering as a matter of economic leadership, rather than as something to pursue because, for example, it threatens a basic resource like water, as the following speaker from DG Research and Innovation argues:

> We will carry on trying to pursue our leadership role in the mitigation of GHGs emissions. Yes, even though we hold only 10 per cent on global emissions . . . *not leadership for the pleasure of leadership, it's because it's in our interest we have to develop these technologies, those industrial sectors, those new ways of living in order to help our competitiveness, growth, and jobs. It's not only climate from our thermometer* point of view. It's in our own interest being in a leadership position.
>
> (Interviewee 9, 16 October 2018)

From these examples, it seems that despite stating that today the work is more horizontal (Interviewee 9, 16 October 2018), *mitigation* is clearly a signifier which is pushed and pulled across the silos' signification spaces, which are themselves an expression of the bureaucratisation of knowledge and organisation of work, an effect of the S2/S1 relationship.

The discussion has aimed to reveal the strictly speaking 'nonsensical character' of the signifier *mitigation* and to challenge its alleged meaning stability by emphasising the role of the real subject of the enunciation and exposing the fractures. In this respect, *mitigation* as such is nonsensical, not because there are no GHG emissions to reduce, or no global warming to tackle, but because the way it is signified in the enunciation reveals each time its partial, incomplete character which we can detect in the inconsistencies that expose the surplus-loss of sense. Nevertheless, we as speaking beings establish, understand, and maintain our climate change relationships by resorting to it as a presupposition of sense, due to the necessary anchoring function of the Master Signifier of competitiveness, accumulation, and growth which is in this case disavowed, and we put knowledge to work through this shared socio-symbolic structure. This is what we mean when we state that we are spoken by language and cannot control all the effects of it. Yet, I demonstrate it is possible to look for disruptive events in which this surplus and/or residue of meaning re-emerges in its force. Indeed, in the next section, the Hysteric subject calls into question the foundation of the hegemonic discourse.

Evidence-based knowledge versus real scientific enquiry: when hysterics defy the dominant discourse

In the policy landscape observed, there are forces attempting to question the hegemonic discourse by introducing elements of disturbance in signification. In this respect, one event that caught my attention was the Postgrowth Conference (see Table A.2, in Appendix). This event was not organised by the EU Parliament as an EU institution. Rather, it was organised by different European party groups that compose the European Parliament. Although the Postgrowth Conference was not exclusively focused on the environment but more widely on further issues of social justice and redistribution of wealth – this way it avoided the reductionism that characterises other political domains – environmental degradation and energy issues were among the topics of discussion. Leaving aside the struggle over the meaning of the signifier *Postgrowth*, which has been criticised by some participants (e.g., the Degrowth activists), this event brought together researchers, activists, and EU representatives from the Commission and the Parliament. Or, to use the words of Molly Scott-Cato, at that time MEP (Greens/EFA), 'a panel *divided between those who bring ideas and people having to deal with the practicability* of putting things into practice' (Molly Scott, 18 September 2018). This event is important not so much for its salience in the EU decision-making process – and as we will see in the next Chapter it had virtually no impact – but because there is an explicit attempt by the speaking subjects-scientists to challenge the hegemonic conversation by questioning what counts as evidence-based knowledge and by refusing the reductionist, rationalised, and instrumental use of science expressed by mathematisation qua modelling and technologisation.

Within this framework, during a session called 'Squaring the energy cycle', chaired by the MEP Florent Marcellesi (Greens/EFA), Prof. Mario Giampietro, Research Professor at Universitat Autònoma de Barcelona (UAB) was invited to speak. On that occasion, he sat at the same panel with Francesco Ferioli, DG Energy Policy Officer in the Economic Analysis Unit – who was in charge of the energy system modelling – and challenged the concept of evidence-based knowledge in response to the question 'Will we be able to satisfy our energy needs?' (sic):

> What I would like to make clear is that *we do not have enough information, a good understanding of the situation in order to be capable of answering this question.* We can say we have a lot of trouble. But what is really important at the moment all the discussion done on how to tackle sustainability *especially in the energy system are based on urban legends or wishful thinking that is very little reliable knowledge claim.* And this is due to the fact, *I think this is a systemic issue, that we want to apply reductionist science, the science done by Newton, to deal with complex systems. A complex system by definition is a system that cannot be measured by numbers in just one way.* By the time you're measuring something you're missing something else. Unless we are capable of expanding a bit more the way we are handling the analysis of energy problems *we are not sure that what we call evidence based is so based.*
>
> (Prof. Mario Giampietro, 18 September 2018)

This extract demonstrates how the subject qua scientist posed an epistemological open-ended question regarding what constitutes evidence-based knowledge, rather than consider it de facto 'evidence-based' as in the case of the EU stakeholder consultation. Thus, what seems to be at play in this case is the refusal of the reductionist and instrumental use of science expressed by modelling and technologisation as exemplified by Lacan's University discourse. Rather, this is contrasted with a view of a more authentic scientific enquiry that needs to think in complex systems to gain a comprehensive understanding of the real, with the difficulties and contradictions that this entails. In this regard, this scientist can be associated with Lacan's Hysteric who challenges the knowledge and validity of the Master, and points to the difference between the two sciences, one that aims at pure rationalisation, and one that 'tries to come to grips with the real, to maintain the difficulties posed by apparent logical and/or physical contradictions' (Fink, 1999, 37).

These events of 'hystericization' defied the unquestioned and untouchable evidence-based nature of the status quo knowledge and demonstrated two points. First, these exemplified how it is theoretically possible to leverage upon these fractures to ideally bring about change. Building on this, they appear to confirm that the real authority of the status quo social bond is in fact disavowed under an alleged spirit of freedom and democracy. However, as exemplified by the story of *energy efficiency* and *renewables* narrated in the next Chapter, these overtly defiant hysteric attempts held little relevance in the EU decision-making process.

Conclusion

This chapter constituted the point of departure for reflecting on the EU's climate action in the period in-between strategies. My aim was to reflect on the nature of climate change knowledge and to show how in fact a seemingly closed discourse can be defied and disrupted to open possibilities for change.

Starting from an analysis of the signifying chain as empirically detected in the field, I claimed that the current EU's climate change mitigation action has been placed in a context of knowledge. Despite being welcomed as objective and evidence-based, this knowledge tends in fact towards rationalisation, bureaucratisation, and valorisation and is put to work in function of command of competitiveness, accumulation, and growth, which is ultimately what constitutes its disavowed yet real Master Signifier, as exemplified by Lacan's University discourse. Empirically, this element of disavowal emerged in the untouchable and unchallenged role accorded to the consumer in this energy transition which cannot be questioned allegedly for the sake of freedom and democracy.

The disavowal of the real authority of the social bond is key to the discussion because it is precisely what makes a performative social act of representation look factual and objective or, as presented in this case, 'evidence-based'. Importantly, the effect of this disavowed anchor of representation commanding the chain of knowledge resulted in climate change and mitigation being signified and constituted mainly along the semantic gravitation centre of energy transition which locked in the climate change's signification space and excluded that from wider considerations about the planet and the ecosystem. Yet, the fact that the intelligibility of reality is fixed by the Master Signifier (S1) qua anchor of representation should not mislead us into thinking that the discourse is a closed structure. In fact, my LDA exposed the unintended effects of the structure on the subjects observed and interviewed providing the examples of signifiers such as *climate change* and *mitigation* themselves, of which I demonstrated the fractures and the inconsistencies. Notably, these gaps embody the *objet petit a* qua ontological excess-residue of signification on which subjects should leverage to work towards a direction of real change. To advance this claim further, the chapter has provided examples of 'hystericization' of the discourse detected in the field. These are exemplified both by the DG Environment actor warning against the environmental consequences of some climate change policies and by some scientists attending the Postgrowth Conference along with EU representatives. At this event, these hysterics overtly challenged what constitutes evidence-based knowledge. These episodes 'disturbed' and defied the assumed and unquestioned evidence-based traits of the status quo knowledge and demonstrated their usefulness on two fronts. First, these demonstrated how it is theoretically possible to leverage upon these fractures to reflect on and ideally bring about real change. Second and building on this, they have indirectly confirmed the element of disavowal concerning the real authority of the status quo social bond.

While this chapter discussed 'knowledge' in general terms, the next chapter will assess how knowledge is translated into concrete policy tools – some of which are in principle desirable – and will investigate the signifiers *energy efficiency* and *renewables* to expose any produced fractures, to show how these are handled in signification by the speaking subjects, and to ultimately assess their transformative potential.

Notes

1 This book follows the work of the 2014–2019 Commission and in this context, the two portfolios of Energy and Climate shared the same Commissioner, who at the time of this study was Miguel Arias Cañete. For context, The Jean Claude Juncker Commission and the Antonio Tajani Parliament have ended their term (2014–2019). These were replaced by the Ursula Von der Leyen Commission (2019–2024), the David Sassoli (2019–2022), and Roberta Metsola (2022–2024) Parliament. New EU parliamentary elections took place in June 2024.
2 Translated from Italian.

5 Climate change mitigation action

The case of *energy efficiency* and *renewables*

Introduction

In this chapter, I continue my investigation of the EU's climate mitigation action in the period referred to as 'in-between strategies' as a formalisation of Lacan's University discourse and present the case of *energy efficiency* and *renewables*. These two policy tools constitute an interesting case study because in principle no one would object to having equipment or devices that use less energy or relying on renewable energy sources rather than on fossil fuel energy sources. Yet, building on the previous discussion about knowledge and its real yet disavowed authority, I zoom in on *energy efficiency* and *renewables* within the 'climate mitigation' + 'energy transition' metonymy to investigate their transformative potential. First, I show that *energy efficiency* and *renewables* as a presupposition of sense mobilise a set of socio-political climate relationships between policy-makers, stakeholders, and citizens around delivering emissions reductions. At the same time, I will emphasise the effects of the signifying chain on the subject and will demonstrate how *energy efficiency* and *renewables* are both spoken in different forms and shapes which are not all necessarily consistent with one another and result in a partial, never-achieved, and deferred meaning across the signifying chain. These inconsistencies are key in the discussion in so far as they embody *objet petit a*, the surplus of meaning in the enunciation upon which subjects could potentially leverage to bring about real change. By exploring the attempts at 'hystericizing' the University and/or Capitalist discourse perpetrated by scientists and academics about the side effects of energy efficiency and about the feasibility of the transition to renewables, I demonstrate that the traumatic points in the discourse do not emerge as an irreversible rupture but are positively integrated into signification and turned into commodified knowledge and into consumption objects. In other words, the acts of 'hystericization' identified do not emerge as disruptive forces capable of producing new significations and real change in the phase of policy-making under observation as every encounter with the fractures of the discourse as a potentially disruptive instance is resisted, that is to say, minimised or downplayed. As a result, the EU's climate mitigation action as embodied by *energy efficiency* and *renewables* appears to be a deceptive transition – the inevitable derivative of the climate + energy metonymy – under the semblance of an efficient and perhaps renewable Master but within

DOI: 10.4324/9781003378228-5

the boundaries of a disavowed S1/S2 power-knowledge relationship. This social fiction, which is what makes access to material reality possible, is supported by the fantasies of efficient products and renewables that promise to cover the impossibility of the discourse and create a continuous sense of paradoxical (dis)satisfaction that does not seem to attain the full enjoyment of the desired climate objectives.

The relevance and desirability of *energy efficiency* and *renewables*

In the previous chapter, I challenged the evidence-based knowledge underpinning the EU's climate change mitigation and disrupted the alleged meaning stability of the signifier *mitigation* with the aim to reveal the unintended effects of the signifying structure on the real speaking subject. These effects emerged in the 'nonsensical' character of the signifier and the visible produced fractures in the discourse. As a reminder, *mitigation* as such has been claimed to be 'nonsensical', not because there are no GHG emissions to curb, or no further global warming to prevent, but because the way it is signified in the enunciation reveals each time its partial, incomplete character which we can detect in the inconsistencies that expose the surplus of sense. Nevertheless, we – as societal actors – establish, understand, and maintain our climate change relationships by resorting to the Other as a presupposition of sense and this is due to the necessary anchoring function of the Master Signifier (in this case of competitiveness, accumulation, and growth) which puts knowledge to work through the shared socio-symbolic structure. Placing the discussion about the EU's climate change mitigation action in the context of fact-based knowledge, whose command is in fact disavowed, poses questions on how to make sense out of policy tools that are key to this general mitigation strategy in relation to their transformative force.

Energy efficiency and *renewables* are key signifiers for delivering emissions reductions both in the EU's medium-term (2030) and long-term goals (2050). The cases of *energy efficiency* and *renewables* – as well as the case of circular economy in the next chapter – are emblematic as they can be regarded as a formalisation of knowledge at work. Their salience lies also in the fact that in principle no one would argue against having equipment or devices that consume less energy or relying on renewable energy sources rather than on fossil fuel energy sources. However, this book is not interested in assessing whether these two measures are going to be enough in bringing about the required GHG emissions reductions – since, as I have highlighted in the previous Chapter, these goals are in line with economic growth – but how *energy efficiency* and *renewables* are signified as climate change mitigation and transition and transformation carriers. The next sections will therefore investigate *energy efficiency* and *renewables*, at the level of their enunciation, to understand what type of transition they are aiming to deliver and whether they carry any rupture potential.

The choice of *energy efficiency* and *renewables* in relation to climate mitigation is not surprising, as these signifiers embody strictly speaking a set of regulatory frameworks but more widely a set of relationships among policy-makers, stakeholders, and citizens that play a vital role in the 2030 Clean energy package. This is exemplified by the Energy Efficiency Directive and the Renewable Energy Directive, as stated by the Commissioner Miguel Arias Cañete, at 'the EU vision for a clean, modern and competitive economy':

> We agreed on *key elements to reduce GHG emissions namely energy efficiency, renewables* governance of climate-energy policies.
>
> (Miguel Arias Cañete, 10 July 2018)

Similarly, their prominent role is highlighted in the 2050 long-term decarbonisation strategy:

> *Energy efficiency measures should play a central role in reaching net-zero greenhouse emissions by 2050* reducing energy consumption by as much as half compared to 2005. Energy efficiency, digitalisation, and home automation, labelling and setting standards have effects that go far beyond the EU as appliances and electronics are imported into the EU or exported to foreign markets, making producers abroad use the EU standards. *Energy efficiency will play a central role in decarbonising industrial processes but much of the reduced energy demand will occur in buildings*, in both the residential and services sectors, which today are responsible for 40% of energy consumption.
>
> (COM (2018) 773 final 8)

The same vital role seems to apply to renewables as well (sic):

> Today the major part of the energy system is based on fossil fuels. All scenarios assessed imply that *by mid century this will change radically with the large-scale electrification of the energy system driven by the deployment of renewables*, be it at the level of end-users or to produce carbon-free fuels and feedstock for the industry. *The clean energy transition would result in an energy system where primary energy supply would largely come from renewable energy sources*, thereby significantly improving security of supply, and fostering domestic jobs. . . . *The large-scale deployment of renewables will lead to the electrification of our economy and to a high degree of decentralisation.* . . . Today, more than half of Europe's electricity supply is free from greenhouse gas emissions. *By 2050, more than 80% of electricity will be coming from renewable energy sources* (increasingly located offshore).
>
> (COM (2018) 773 final, 8–9)

From these extracts we understand that *energy efficiency* and *renewables* are and will be the key tools for delivering emissions reductions both in the medium term and in the long term. Again, I am not interested in assessing whether the two measures are going to be sufficient in bringing about the required GHG emissions reductions but rather to analyse their rupture potential. In fact, according to the EU Commission these reductions were estimated at only 60 per cent, against the 80–95 per cent required, as specified in the long-term strategy Communication: 'The policies put in place today will have a continued impact after 2030 and will therefore already go a long way, with projected emissions reductions of around 60% by 2050. This is, however, not sufficient for the EU to contribute to the Paris Agreement's temperature goals' (COM (2018) 773 final, 5), and this explains the advancements and the targets increase outlined in Chapter 2, notably the 'Fit for 55 Package', which aims to revise existing legislation and set more ambitious targets for 2030 with the aim to ultimately bring this process back on track with the 2050 Paris Agreement goals.

To state that *energy efficiency* and *renewables* are 'knowledge' that performs climate mitigation action is emblematic since, in principle, no one would argue against having equipment or devices that consume less energy or relying on renewable energy sources rather than on fossil fuels energy sources. Hence, *energy efficiency* appears a good neutral tool at the service of climate change mitigation, especially if considered in relation to the waste of 'energy inefficiency'. Similarly, renewables as a regenerative and clean form of energy as opposed to the dirty and finite fossil fuels combustion are no doubt desirable. As a result, at this stage of the discussion, the point is to understand what *energy efficiency* and *renewables* are as produced in the signification via the subject of the enunciation and reflect on the type of transition or transformation they deliver within the context of (a given understanding of) climate change mitigation action.

Energy efficiency qua lacking knowledge

In this section, I am going to apply a LDA and disrupt how *energy efficiency*, which is something intangible, is being signified by the subject of the enunciation. We will see that energy efficiency as a presupposition of sense mobilises and keeps together a set of socio-political climate relationships between policy-makers, stakeholders at large and citizens to deliver emissions reductions. At the same time, it takes different significations, which are not necessarily consistent with one another. This can be explained via a Lacanian approach in that such inconsistencies embody *objet petit a*, the surplus and loss of meaning in the subject's enunciation, emphasising the partiality of signification, while full meaning is deferred across the signifying chain.

First, behind the *energy efficiency* signifier lies a measurement. An EU Commission official from DG Energy during an interview clarified that it would be more appropriate to speak of energy savings:

> The equation regulating energy efficiency is related to the *increased energy saving per unit of production*, it is the savings you obtain compared to a situation in which that measure/technology is absent.
>
> (Interviewee 6a, 7 November 2018)

The EU official explains that this equation and the relative function drawn on a whiteboard do not tell us anything if, for example, the population grows. We can deduce then that if the population grows and we need more buildings, however efficient they might be, the energy demand grows, and this increment is not embedded in the *energy efficiency* equation.

Second, when energy efficiency occurs along with Renewables and Governance, it refers to the legislative acts named after it, as illustrated by the following statement from the MEP Sean Kelly (EPP) at the 'EU vision for a clean, modern and competitive economy':

> In terms of moving to a low-carbon system it is vital that we think long-term, imagine how the EU economy would look in 2050 and beyond. . . . First, from the outset it is important to remind ourselves that we have an obligation to meet the commitments under the Paris Agreement and ensure that the required emission cuts are made. That is not something that is optional. Sustained and consistent action on climate change is a must. The only debate to be had is to consider how we do it not if we do it. *In the past few weeks, the Parliament and the Council reached an agreement on three key elements of the Clean energy package which was pushed so much so and so effectively by Commissioner Cañete. The Renewable Energy Directive, the Energy Efficiency Directive, and the Governance Regulation.*
>
> (Sean Kelly, 10 July 2018)

In this case, these legislative tools are claimed to be the means for achieving the targets of the 2030 Clean energy package. Thus, in the linguistic representation, the Energy Efficiency qua Directive becomes a metaphor for the concept of efficiency (energy saving).

Third, the International Energy Agency (IEA) was invited to speak at the stakeholder consultation 'the EU vision for a clean, modern and competitive economy'. During the Session 'Cost efficient ways for achieving decarbonization' (10 July 2018), its representative, David Turk, Acting Director, Sustainability, Technology, and Outlooks presented a slide about the existing gap between climate commitments (Paris Agreement and SDGs) and the current mitigation trends. This gap is filled in the graph with percentages of technologies as following: Energy efficiency 44 per cent; Renewables 36 per cent; Fuel switching 2 per cent; Nuclear 6 per cent; CCS 9 per cent; and Other 2 per cent. The slide caption cites:

> A wide variety of technologies are necessary to meet goals, *with energy efficiency and renewables playing lead roles.*
>
> (David Turk, 10 July 2018)

In this case *energy efficiency* appears as an item among available technologies, along with CCS, nuclear (technology as well as an energy source) and renewables (technology and energy sources), for achieving GHGs reductions. Again, this is another metaphor, and it stands for technology in itself. In other words, it is not meant to encompass all technologies that should perform 'efficiently', such as, for instance, the efficiency of modern solar panels compared to ten years ago, or the efficiency of nuclear reactors compared to renewable technologies.

Fourth, in the following extract, our interviewee comes up as a free association as to what *energy efficiency* does not represent:

> Something is right then, our policy is called energy efficiency, our goal is an absolute low level of energy consumption, so *if efficiency means how much how improved the quality of output you get for a given input of energy, then our policy is not enough*, it's energy saving policy.
>
> (Interviewee 8, 14 November 2018)

Fifth, if *energy efficiency* means savings, it also means savings in economic terms, as the following DG Energy official explains:

> Energy efficiency is not only about CO_2 emissions, but also about *savings. It makes sense in the economic sense.*
>
> (Interviewee 7, 13 November 2018)

It can be argued that when 'energy efficiency' slides into 'energy savings', it is possible that the mentioned savings are not necessarily the needed energy savings, but economic savings. As a result, when *energy efficiency* is juxtaposed with cost-efficiency, speaking only of 'efficiency' makes it unclear whether efficiency is an attribute of energy or of costs. Of course, if energy efficiency can be achieved in a cost-efficient manner, this is not necessarily a negative aspect, but the cost-efficient logic does not necessarily lead to a reduction in energy volume as we will see later in the Chapter.

Finally, the possible understandings around *energy efficiency* can be further expanded. In the following extract, the DG Energy's Director General Dominique Ristori gave a speech at the 'EU vision for a clean, modern and competitive economy'. During the session 'Cost-efficient ways for achieving a post-carbon European economy', he stated:

> Our vision is a vision of a multi-sectorial approach, *beyond the energy sector, beyond the power sector.* But examining *also the situation of the industry sector, of the mobility and transport sector, of agriculture.* This is fundamental in order to open the road regarding the decarbonisation of the whole economy.
>
> (Dominique Ristori, 10 July 2018)

This case shows that at the EU level, perhaps as a result of working in silos, they speak of the 'energy sector' when they mean the power sector, that is, electricity and electrification. As a result, industry, transport, and buildings which in turn require 'energy' from any source, to which the concept of 'efficiency' qua savings should equally apply, constitute separate areas. This tends to generate overlapping or confusing juxtapositions in representations and categorisations. For example, if transport (as a different sector from the energy sector, it is also formalised in two different Commission DGs) is electrified, such as in the case of electric vehicles

(EVs), it becomes unclear whether EVs are counted in the energy sector qua power sector, or in the transport sector.

In these examples, we can appreciate how speaking subjects cannot escape the Symbolic and are structured in relation to the Other as the 'locus in which is situated the chain of the signifier that governs whatever may be made present of the subject' (Lacan et al., 1998, 203; Lacan, 1964). In this respect, *energy efficiency* as a presupposition of sense mobilises and keeps together a set of socio-political climate relationships between policy-makers, stakeholders, and citizens around it to deliver emission reductions but takes different meaning forms and shapes – which are not necessarily consistent with one another – depending on how it is unconsciously signified in the (split) subject's enunciation because of the irreducible cut that language generates within subjectivity (Lacan, 1964; Lacan et al., 1998, 204–205). The signification of *energy efficiency* is thus always partial, never-achieved, and deferred across the signifying chain, as an effect of this ontological lack as excess of meaning, that is *objet petit a*, and the same logic can be applied to renewables.

Renewables qua lacking knowledge

As opposed to the abstractness and intangibleness of *energy efficiency*, *renewables* recall something that we can picture in our minds, whether in the form of wind, sun, water, or the associated technology, such as turbines and solar panels. Thus, *energy efficiency* has a more tangible nature than *energy efficiency*. However, as with the case of *energy efficiency*, *renewables* as well are spoken of as a presupposition of sense that holds a set of socio-political climate relationships between policy-makers, stakeholders, and citizens, but it is articulated in different forms and shapes depending on how it is signified in the enunciation, and most often inconsistently.

First, as with the case of *energy efficiency*, when the signifier *renewables* appears next to *energy efficiency* and *governance*, it is meant as a legislative act, that is the Renewable Energy Directive, as the case of Commissioner Cañete at 'the EU vision for a clean, modern and competitive economy' illustrates:

> We agreed on *key elements to reduce GHG emissions namely energy efficiency, renewables governance of climate-energy policies.*
>
> (Miguel Arias Cañete, 10 July 2018)

As with the case of energy efficiency, the legislative act becomes a metaphor for renewable energy or associated technologies.

Second, as we saw in the case of *energy efficiency*, quantifying targets is what mainly drives signification. Accordingly, a DG Energy official explained that behind *renewables* lies a measurement. However, such target calculations consider the industrial capacity of Member states, rather than their natural capacity:

> This is when the target is set . . . there is a formula. *At the moment I don't know what the formula takes into account. I think it takes into account the*

natural potential as well, but I need to check it with my colleagues, we can do it afterwards.

(Interviewee 6, 07 November 2018)

A co-worker within the same Unit of DG Energy confirms on an informal conversation:

Mainly industrial capacity, infrastructure, GDP rather than natural potential.
(Interviewee 6a, 07 November 2018)

Third, when speaking of *renewables*, this can mean both the type of energy sources (i.e., solar, wind, etc.) and renewables technologies (i.e., solar PVs, wind turbines). Thus, speaking of renewables can either refer to their natural capacity for energy generation or their industrial capacity. Therefore, a distinction must be drawn, in that the energy source as natural potential is different from the technology that involves a product's life cycle, from the extraction of raw materials to the product's disposal and replacement. As a speaker from DG Energy explains (sic), some countries might have the industrial capacity but not the natural potential:

Here, the Netherlands you can see a bit low[1] which should be where to they are, so they really need to take measures. . . . Yes, but if you look at how low the renewables was because France and other *countries with mountains have lots of hydro. This is not an option for the Netherlands* [. . .]. First, the Netherlands are quite densely populated. *They don't have hydro which is a traditional source of renewable energy. They just don't have the natural potential for that.*
(Interviewee 6, 07 November 2018)

Fourth, from the perspective of the climate lead DGs, that is DG Clima and to a good extent DG Energy, *renewables* appear to be signified in isolation from the ecosystem and biodiversity considerations. This is a perspective that is challenged by DG Environment, which highlights that renewables are not necessarily the ultimate solution to climate mitigation:

But this is perhaps our role, to say be careful that at present *the majority of renewable energy comes from biomass. If we carry on this way, with bioeconomy etc. we won't have biomass eventually and we'll destroy the ecosystem services and biodiversity.* Thus, this is our problem. But there are also many synergies, I started with the negative aspect, as adaptation it is clear, water retention, flood protection, reduce the temperature when there are trees etc., the carbon sink in terms of mitigation, thus there are many synergies, and we defend those because we know that biodiversity is in danger. *If now climate change is the greatest problem of humanity, the second is probably biodiversity loss.*
(Interviewee 14, 29 November 2018)[2]

Finally, and building on the previous argument, the silos' approach to work is far from being overcome. This aspect emerges in the words of an interviewee from DG

Energy who takes for granted the transition to renewables, in isolation from other policy considerations:

> It's true I mean *I'm in a renewables bubble, so I don't know the discussion taking place within the coal lobby.*
>
> (Interviewee 6, 07 November 2018)

Similarly, another EU official from the same DG states:

> *The guys from renewables would like more but clash with other departments. . . .* That's what our renewables colleagues would like to have very high ambitions in terms of renewables. Then you might have for instance . . . *In the renewables people want higher targets than we have in DG Grow. DG Ecofin says this is going to be very expensive,* how are we going to finance that? This kind of technical questions from their perspective, so the renewables is not the best question because it's more the Member States who would say whether it's feasible or not. So, there is a tendency within the Commission to agree on ambitious targets.
>
> (Interviewee 7, 13 November 2018)

In conclusion, as with the case of *energy efficiency*, *renewables* as presupposition of sense mobilises and holds together a set of socio-political relationships between policy-makers, stakeholders at large, and citizens. The cases of *energy efficiency* and *renewables* so far illustrated are, I would suggest, exemplary because as targets, modelling, legislative acts, and technologies they stand for knowledge as a quantifiable entity at work, the type of knowledge indicated by Lacan's University discourse asserting itself as neutral, measurable, quantifiable, and bureaucratised, rational, and objective but disavowing its own authority. Moreover, it can be suggested that *energy efficiency* and *renewables* are signified within this existing battery of signifiers as in this climate mitigation + energy transition nexus and are ultimately its inevitable derivatives. Yet, notwithstanding this signification closure, individually each signifier, that is, *energy efficiency* or *renewables*, cannot confer the meaning that the entire chain offers, but they are signified in different meaning forms and shapes, as a legislative act, as an equation, as a technology, as economic savings, as a type of energy source at different times of enunciations, which are not necessarily consistent with each other. These points of rupture demonstrate that the discourse seems locked, whereas is in fact open and fractured and signification is never-achieved and often inconsistent because the act of speaking introduces a cut within subjectivity. The way these traumatic points are handled by the speaking subjects help assess what type of transition and change *energy efficiency* and *renewables* can deliver in this climate social bond.

Energy efficiency and *renewables*: a Lacanian fantasy?

Disrupting and disorganising the spoken texts of *energy efficiency* and *renewables* does not aim to dig up any hidden meaning and establish an absolute truth of what they really mean. In fact, starting from the signifier and its centrality in the formation of subjectivity, I have illustrated that *energy efficiency* and *renewables* are all that is being spoken, including the attendant inconsistencies and gaps, that which escapes language. This is precisely what makes the discourse inherently inconsistent, never complete, and always deferred to other signifiers. In this leftover, which represents the gap between what the subject wants to say and what the subject says, lies that 'lack' that sets desire in motion, which Lacan identified as *objet a* – the intangible non-linguistic object-cause of desire, as excess-loss to the discourse (the Real). It is this constant quest for full representation, embodied by this object, that keeps discourse and subject together (Solomon, 2015, 66, 127) and we can approximately compare this *objet petit a* qua lack to a never-achieved state of wholeness that we still aim to (re)capture, such as a harmonious and safe planet, a growing prosperous environment, abundance of energy supply, efficient and infinite clean resources.

Moreover, while this does not mean that all the EU representatives are necessarily convinced by or consciously aware of this discourse, they nevertheless appear to desire to believe it, as in the case of this interviewee from DG Energy:

> I am very concerned about climate change, and I am willing to . . . I mean, *I have views as a private person, but I think that as a person I'm also very concerned about climate change and I'm interested in how we're going to get to avoid catastrophic global warming* by 2050 and the work that the Commission is doing, the long-term strategy, is interesting in that way. *I don't think the modelling that we're doing would be the end of the story because at least you start having some stories* about how you can get to 1.5 degrees, *stories in the sense of pieces of modelling which add up on some basis which describes worlds* you know. *And they have an x amount of electric cars, an x amount of CCS or whatever you have in them,* and these are worlds that would get to 1.5 degrees. We've got better and better at modelling these things.
>
> (Interviewee 8, 14 November 2018)

As the extract indicates, the subject above is caught in a net of signifiers such as stories, modelling, a given number of electric cars, CCS, but a split is produced between personal considerations and the EU representative as whom they speak. While the subject seems to be aware that this discourse is full of 'holes', points in which the field of representation breaks down, there is an attempt to disavow personal considerations as an EU representative and embrace and turn the traumatic points of the socio-symbolic Other into a positive feature. This can be seen in the fact that the speaker desires to believe 'some stories', under the forms of commodified knowledge 'pieces of modelling' or consumption objects 'x amount of electric cars' (Interviewee 8, 14 November 2018). This way having pieces of modelling as well as a given number of electric cars represent a Lacanian fantasy, a reassuring symbolic construction that sutures the subject, promising to eliminate the lack and anxiety in the subject and to cover the impossibility of full representation by achieving a given desired object. However, every fantasy formation is articulated around the *objet petit a* (object qua lack), and this missed realisation is exactly what sustains the promise of fullness (Fletcher and Rammelt, 2017; Stavrakakis, 1999, 46–51) fuelling a process of *jouissance* and keeping the subject suspended by temporary manifestations of (painful) enjoyment (Mandelbaum, 2022).

In summary, the subjects observed and interviewed appeared to be split between these powerful linguistic structures defining them, and their desire to reach the ultimate transition and full satisfaction that is always deferred. In this respect, in the pursuit of achieving climate mitigation as a form of full representation and ultimate enjoyment, the EU mobilised its knowledge apparatus and from the desiring position of having an effective climate policy these split, fraught subjectivities – as workers and consumers themselves – set 'knowledge' to work with the available signifiers. These subjectivities are therefore the EU bureaucrats themselves as employees caught in their chain of consensus-seeking practices as well as stakeholders negotiating and lobbying their targets with policy-makers. As a result, they established the targets, formulated impact assessments and cost-benefit analysis, conducted official and unofficial consultations at the EU level and international levels, engaged with modelling activities, configured scenarios, released official policy documents, and funded technology innovation and research. The split subjectivities also become the new generation of graduates ready for the green job market, as well as all the new citizens as consumers put at the centre of this climate and energy transition who will participate and enact their smart consumption practices in a flexible decentralised smart energy market. In other words, these split subjectivities seem to be the product of the knowledge-based social link.

Yet, within this framework, these subjectivities clang to the socio-symbolic construct of fantasy of 'having some stories' (Interviewee 8, 14 November 2018), meant as pieces of modelling, which gave them some reassurance that climate change is being tackled. At the same time, these subjectivities seem trapped in their closed circuit of commodified knowledge and objects, which in turn resulted in an endless (dis)satisfaction and illusionary sense of plenitude, what we call *jouissance*. This manifests as a need to resort to more sophisticated modelling, greater bureaucratic organisation and policy integration at all levels of climate governance as well as a need to produce and consume more efficiently, that is, through efficient buildings, effective heating and cooling systems, efficient cars, efficient industrial processes, and therefore it can be concluded that *objet petit a* as surplus-excess to the discourse is positively integrated into signification, with its rupture potential being severely reduced if not nearly neutralised.

For this reason, the EU's climate mitigation social bond exemplified by *energy efficiency* and *renewables* can be regarded as a formalisation of Lacan's University discourse, because allegedly neutral scientific knowledge (S2), which in fact disavows its real authority, attempts to control and tame *objet a*, the surplus of meaning perceived as a loss that sets desire in motion. *Objet petit a* and its related *jouissance* emerged in all the ways in which *energy efficiency* and *renewables* are enunciated and integrated into signification by turning 'lack' into a consumption object. When this object is integrated into signification, it loses significantly its disturbing, traumatic, and therefore transformative potential (Feldner and Vighi,

2015, 93; Wright, 2016, 142) and knowledge becomes the vehicle of production, commodification, and diffusion of *jouissance* (Wright, 2016, 138–142). At the same time, as surplus *'jouissance'* has become calculable (Lacan, 2007, 177) as exemplified by the EU's heavy reliance on targets, it can also be a formalisation of the Capitalist discourse and its exploitation of the structure of the desiring lacking subject as a means of endlessly reproducing itself (Lacan, 1973, 94). As a reminder, in this fifth discourse, the alienated and powerless worker-consumer ($) is in the agent position and sets knowledge S2 in motion with the aim to fill their lack through the (concealed) Master Signifier S1 – the same command of the University discourse – and demands objects of enjoyment that fuel this lack rather than satisfy it. It is, in this discourse, that we can appreciate how surplus value functions as surplus *jouissance*.

It is possible however to look for disruptive events in which this residue re-emerges in its force: in the next section, we can observe how the Hysteric subject calls into question the foundation of the discourse by explicitly challenging the knowledge in which *energy efficiency* and *renewables* are spoken.

The symptom of *energy efficiency* and *renewables*

Attending the Postgrowth Conference – the event which brought together researchers, activists, and EU representatives from the Commission and the Parliament mentioned in the previous chapter – made it possible to expose the ruptures of the discourse through *energy efficiency* and *renewables* through the 'hystericization' acts of some of its speaking subjects. These can be detected in the disavowal of what is known as the 'rebound effect', that is the collateral of the climate discourse as a discourse of knowledge. Moreover, these ruptures emerge when issues of energy volumes are introduced and opposed to an idea of sufficiency, associated with a signifying chain of frugality. As it will emerge, these defiant subjects in the field refuse the evidence-based knowledge of the EU Commission in that *energy efficiency* does not necessarily mean a reduction in energy volume.

During the panel 'Squaring the Energy Cycle', Dr. Grégoire Wallenborn, from Université Libre de Bruxelles (ULB) zooms in on energy efficiency:

> What matters is the *volume of energy*. If we want to reduce *energy consumption, we have to reduce growth. . . . Energy efficiency is often considered as the solution to all energy problems, and I think that this is really misleading.* I am trying to explain to you what is energy efficiency. *Efficiency is the same as productivity. Efficiency is a ratio between out and input and when you consider energy efficiency you will consider the same activity with less input, less energy. But if the idea is to optimize, maximise efficiency you can also just with the same input have more output and more activities.* So, if policies are aiming at improving energy efficiency, if efficiency is just an end in itself it leads to just more energy productivity. And that explains why you don't observe energy reduction while energy efficiency is rising. *So, I really think the rebound effects are systemic, they are difficult for modelling because it*

is a complex system. if you ask the question: if you save energy somewhere, where will it be used? You can imagine if you save energy efficiency that you'll have at some point in some activity but if you have infrastructure and markets which allow for its redistribution you won't have energy reduction. *So, the rebounds are the reinterpretation that we can increase the number of activities with the same energy and it's really what we see today.*

(Dr. Grégoire Wallenborn, 18 September 2018)

Although it has already been highlighted that energy efficiency is a ratio between product output and energy input, at this event the fractures in the discourse are explicitly exposed. For the first time, we hear someone speaking of *volume* of energy. Let us consider again *energy efficiency* as an empty signifier. Improvements in efficiency create energy savings, a 'surplus-energy' we might call it. If its meaning comes from the interplay of signifiers having accumulation as their anchoring point, these savings are most likely to be re-invested and re-absorbed into the market and infrastructures with a negligible mitigating effect, to the point that we can establish an equivalence between the chain and its anchoring point. The surplus energy then becomes nothing but an economic surplus value in the strict sense. At present, *energy efficiency* virtually equates with productivity, which does not automatically translate into a decrease in energy volume and the reason for that is the so-called rebound effect. This is the 'surplus-energy' gained with efficiency improvements that is re-invested.

Conversely, if the amount of energy saved through efficiency measures is indeed saved, and if *efficiency* is attached to a signifying chain of *energy volume, sufficiency*, and *planet boundaries*, then alternative significations and new subjectivities other than the worker-consumer could be produced. Theoretically speaking, it is this new signification and new subject that can be associated with Lacan's discourse of the Analyst that would ideally bring about the systemic change and the new societal relations DG RTD interviewee expressed: 'Now we know that we run for a completely systemic change, a new society a new system a new behaviour, new social relations (Interviewee 9, 16 November 2018)'.

Interestingly, the existence of the 'rebound effect' as a side effect of energy-efficiency measures has been repressed for some time, as the chair of the 'Squaring the energy cycle' MEP Florent Marcellesi explained:

And then we have the rebound effect, I think it's quite important that you mentioned that because it's something that I said several times to the EU Commission rebound effect *and never had an answer about that*, because I was the shadow reporter for the greens on lots of energy issues, and *that was something that I don't understand that we don't include in the reflections.*

(Florent Marcellesi, 18 September 2018)

In fact, during a session called 'Energy sufficiency' (and we should note that the signifier is now sufficiency not efficiency) with Fulvia Raffaëlli – DG Grow, Head of

Unit responsible for Clean Technologies and Products – and Philippe Tulkens – DG Research & Innovation, Energy Directorate, Deputy Head of Unit – Prof. Blake Alcott from Cambridge University expanded on the rebound effect. The Q&A session after presentations leaves a question (remained unanswered) to DG Grow:

> One question is to Fulvia, could you divert some research money to *a new rebound study*, a scoping study whatever it is called . . . and go back to the rebound question from in DG grow. You talked about new indicators or something and really go into this question of if efficiency changes then we increase growth I think we all agree on this. *I think that now the people in the IEA* (International Energy Agency) *say right now that rebound is wonderful because it increases prosperity; and this is hilarious to me because it used to be denied and now it is big and the bigger it is the better and it's very good for GDP, good for affluence but that's not good for degrowing throughputs.*
>
> (Prof. Blake Alcott, 19 September 2018)

In the above extract, the speaker insinuated some resistance to the very existence of this *rebound effect*. In fact, during my interviews, the rebound effect never came up spontaneously, unless explicitly mentioned. This seems to suggest that the rebound effect is a collateral of the climate social bond as knowledge was previously disavowed and when it re-emerged repeatedly, in a 'return of the repressed' (Bettini, 2019; Hook, 2013; Lacan, 1988), its traumatic force had been instead embraced and turned into something positive and exchangeable. As a result, this upholding of the rebound effect as a positive feature is confirmed by the words of the Commission, as this extract from DG Energy suggests, shortly after I mentioned the rebound effect:

> Of course, *there's a rebound effect, because we save people money* and you know our policy makes economic sense, they have the reasons *that the economic benefit is the rebound effect. The benefit of our policy is split between energy saving and economic growth. You know if you had no rebound effect you can have more energy savings but no more economic growth.* Put it another way, if we ban vacuum cleaner which would be crazy to buy, people who buy vacuum cleaners save more money in their pocket because *they would pay more on a vacuum cleaner, but they spend less on electricity, they do something with that money and that's the rebound effect.*
>
> (Interviewee 8, 14 November 2018)

This seems to confirm what has been discussed earlier in this chapter when it has been emphasised that the cost-efficient logic regarding energy efficiency does not necessarily lean towards a reduction in energy volume: 'if you had no rebound effect you can have more energy savings but no more economic growth' (Interviewee 8, 14 November 2018). So far, the emphasis has been placed on an alternative understanding of *energy efficiency*, where we understand that it is the same as productivity and that the rebound effect is turned into a positive feature of the social order, with a potentially negligible effect on our climate mitigation objectives. What happens then in the case of *renewables?*

Renewables are taken for granted as a necessary part of the energy transition. However, as previously mentioned, *renewables* refer both to the natural energy source available and the technology and infrastructures needed to produce renewable energy. In the following section, the hegemonic understanding of renewables is challenged by issues of source intermittency, technicality of the grid, storage scaling, and amount of electricity for end-users. I argue that renewables as climate knowledge expose the subjects to the discourse's fractures more than energy efficiency, insofar as they question the very same mass transition to renewable energy. At the same time, although an event like the Postgrowth Conference challenged the system of knowledge governing the current EU's climate mitigation action, every traumatic encounter of the lack with its liberating potential is resisted by the EU representatives who embody the hegemonic discourse. I show empirically that its relevance and potentially boundary-breaking character are minimised and downplayed in the official decision-making process environment.

At the session 'Squaring the energy circle', Prof. Mario Giampietro (UAB) speaking next to DG Energy's Ferioli explains that the problem of renewable energy lies in its nature as an intermittent source, what we call natural capacity. The speaker uses the jargon 'intermittent', to explain that it might be available when not needed and vice versa and uses a water metaphor to explain the use of the grid.

This understanding is juxtaposed with that of renewable as a technology, the industrial capacity – the grid that should generate electricity within the context of the structural flaws in the German energy transition plan called Energiewende (sic):

Electricity is a form of energy that is different from other energy. It is mechanical energy. It's moving in pipes, like water. If you have a grid, you have something a baseload producing 400 units of water and users having 400. So, if the user uses more water, you have to out more water in the pipes otherwise the system stops working. So, you have what is called peakers, you have turbines that put the additional water. If the users stop using water, then you have to take out the peaker. Okay this is the way the gird works. *So, what happens if we put intermittency? You put surprise supply, at a certain point you start putting water into the pipes and what happens? If you're not regulating the pipes break down.* If you start putting intermittent in a grid you are generating a lot of trouble. *So, if you're using a base loader nuclear plant or coal plant this is using almost all the time. All the electricity that is produced is used. Peaker, the same.* If you use the peaker when it's needed all the electricity produced is used. *If you're using an intermittent first of all you're using the intermittent very little when there's wind. Especially solar PV in Germany is a very bad idea you use about 10 per cent probably. What happens is that not only you're producing less energy per unit of capital but also maybe when you're producing energy maybe it's not used* because they're producing it not necessarily when it's needed and you're not producing it when it's needed. . . . So, if we don't have a way of having a storage per se the idea of starting investing billions, hundreds of billions is not particularly good.

(Prof. Mario Giampietro, 18 September 2018)

Thus, we understand that the transition to renewables is less smooth than it seems. In the absence of storage, issues arise for three main reasons. First, they arise for technical reasons due to how grids work. Second, they emerge for the intermittent nature of the energy source. Finally, issues arise in terms of the actual electricity amount available to the end-user. Even if we have storage, such as batteries, a scaling issue emerges (sic):

> If Tokyo remains one day without intermittent . . . you will require 600 gigawatt/hour . . . and if the largest battery station they are expected in 2021 will be 400 megawatts . . . , *we 're talking about that we will need 1500 stations, the largest in the world, just to have a back-up for Tokyo for one day.* And if you look at the batteries of the cars and still for 2 million cars, we're already in the order of 160 gigawatt/hour, . . . if you put the cars in Europe that would be an outrageous number of batteries. Let alone the batteries required for backing up the grid. . . . What is the problem? That maybe you have to renew it every 3–4 years or 5 years. Can you imagine this type of things? *Hundreds of thousands of things to be renewed every 3–4 years how much will it cost? Do we have lithium? What are the environmental impacts of recycling all these things? Nobody knows.* I'm not saying it's not possible or that we should not do it. But what I'm saying is that this something that will not happen in the next 10 years, if you're listening in 2030–40 everything is solved.
>
> (Prof. Mario Giampietro, 18 September 2018)

This suggests that there is a huge problem of scaling in terms of renewable technology, battery storage and life cycle, in that they follow a material flow as any other commodity, from its extraction to its disposal or recycling. In this respect, building renewable technologies and infrastructures might be good for the modernisation of industry and the economy but not for climate objectives, as the same speaker as scientist observes:

> Can you go back to the slide number 4 of the quick decarbonisation (referring to the presentation of DG energy Ferioli, Ed.). This to me is *fascinating I don't' know how the models are done* but basically we are saying that in 20 years we will have 40 times more power capacity in terms of . . . *who is building all these things? Because if we are doing this using fossil energy, we will have an increase in the emissions not a decrease. So,* I don't know how the model is done. But if we want to have some projects about how much energy will be required to be all the power capacity and consuming capacity that will be required for this substitution, I'm not sure how reliable this is. *This is why I always say it's better to use story telling rather than models, because models are tricky.*
>
> (Prof. Mario Giampietro, 18 September 2018)

If these renewable technologies are built by burning fossil fuels, then the model reveals its limits in the real world yet his interlocutor – the DG Energy representative – never answered the question about how the model is built. Finally, the problem with biofuels is raised by the scientist-hysteric subject (sic):

> Biofuels, this is my favourite: *fossil energy was about using oil to save labour and land. Biofuels is about using labour and land to save fossil energy.* That is, I mean we missed what happened in this planet over the last 200 years. . . . You can see the on the right you see the density of supply. Please note that this is a logarithmic scale, so it's 10 times more. *We can feed cities and all the industrial use because the density at which we are producing energy with fossil energy is in order of magnitude higher. With biomass is much lower. If we we're using biofuel it's even lower* because we're spending energy to convert biomass in liquid fuels.
>
> (Prof. Mario Giampietro, 18 September 2018)

The fractures and traumatic points of the discourse are exposed and challenged explicitly in the case of renewables as well as in the case of energy efficiency. In a chain of signification in which the total volume of energy as the anchoring point, we hear someone addressing the problem of variability – or intermittency – and the technicality (the materiality of the grids) that make the transition less smooth than the EU policies make it seem. Perhaps, *renewables* as climate knowledge expose the subjects to the traumatic points of the discourse even more explicitly than energy efficiency. More precisely, in the case of *energy efficiency*, the negative can be disavowed and integrated successfully into signification via the appraisal of the rebound effect, which is good for economic growth. Instead, *renewables* expose the discourse to all its limits in terms of infrastructure, variability, storage scaling, and final amount of electricity usage, which ultimately question the very same mass transition to renewable energy.

Although an event like the Postgrowth Conference had a clear intent to challenge the University discourse, that is the system of knowledge governing the current EU's climate mitigation action, it should not mislead us into drawing the wrong conclusions. In this current phase of policy-making, the repressed returns in language but every traumatic encounter with the limits of the discourse as transformative force is resisted. This event was in fact not an official EU event but organised at EU premises by a few political groups within the EU Parliament.[3] In fact, its salience and potentially boundary-breaking character are minimised and downplayed in the official climate policy-making environment. For example, I have shown the post-growth leaflet to a speaker in DG Clima:

> *I don't know* what the event was.
>
> (Interviewee 4, 12 October 2018)

After they looked at the flyer of the event:

> So, 5 political groups, the unions, NGOs, it's unions, NGOs and political parties *all discussing basically how to grow.* I mean, I have not been there but looking at it *this is a lot about what is the definition of sustainable growth.*
>
> (Interviewee 4, 12 October 2018)

As we can see, an ambiguous floating signifier like *postgrowth* is instinctively associated and anchored to 'how to grow' and the 'sustainable growth' representation.

During another interview with DG Energy, I asked why the Commission has decided to take part in something like a Postgrowth Conference. The speaker seems first open to hearing the 'Postgrowth' perspective which they find intellectually interesting. The speaker also justifies the Commission's participation on the grounds that the invitation came from an MEP, so the Commission would normally accept to participate purely as part of their representative role:

> *I find it intellectually interesting,* but if you set that aside whether has to go a speaker or something we normally say yes. You know, at least our presumption is that yes and if a MEP then our presumption is to say yes *immediately because that's part of our role . . . .* I mean it's just that the postgrowth movement *is not something I really interacted with.*
>
> (Interviewee 8, 14 November 2018)

In Chapter 3, I maintained that for Lacan authentic scientific enquiry is associated with the Hysteric's discourse (Lacan et al., 1990, 19; Lacan, 2007, 23) rather than with the University discourse. In the above empirical exemplifications, the speaking subjects above took the agent position and called into question the dominant power-knowledge relationship (S2/S1) governing *energy efficiency* and *renewables*; they can thus be regarded as hysterics insofar as they opposed to a 'science' that does not cover contradictions and paradoxes of knowledge with a 'science' that works for the Master Signifier as in the University discourse. At the same time, the extracts above suggested that the effects of a subject bringing overtly disruptive elements into the scene within a given social bond are conditioned by the law of the dominant existing social bond and therefore resisted. This confirms the argument raised by Fink (1999) who claims that the hysteric can function within the University discourse, but what changes is his or her efficacy. This occurs because the possible effects, shortcomings included, are those allowed by and pertaining to that specific discourse (Fink, 1999, 30).

In conclusion, this exercise therefore is not a critique of *energy efficiency* or *renewables* – as an EU Directive, as an equation, as a technology – in itself. The critique addresses how *energy efficiency* and *renewables* are spoken of by its representatives in relation to delivering an ambitious climate change mitigation strategy and in relation to any transformative potential. What I intended to show with this chapter is that we cannot think of *energy efficiency* or *renewables* as a presupposition of sense that automatically delivers the desired and required reductions. From the analysis conducted, and in relation to the period 'in-between strategies', we are facing a case of signifiers that are well anchored to a powerful and dominant signifier from which they struggle to detach, that is to say, a chain of allegedly neutral, rational and valorised knowledge – which is in line with Lacan's University and Capitalist discourses. From these social bonds, deeply fraught inconsistent subjectivities emerged in their pursuit of an effective climate action and desired 'system change'. As a result, the energy transition qua climate action appears to be an illusionary change supported by the taming of *objet petit a*, that is, the constant commodification and thus integration into the signification of its traumatic points, as illustrated in the case of the recent enthusiasm around the previously denied rebound effect. This ultimately hinders the rupture potential of the discourse and the possibility to bring real change. Rather, this illusionary change is supported at the level of the Symbolic by constructions, that we call fantasies, of efficient products and renewable energy that promise to cover the impossibility of the discourse, whereas at the level of the Real, this functions as surplus-*jouissance*, as a paradoxical (dis)satisfaction that accompanies every missed realisation and ultimate achievement.

This transition looks therefore more like a redistribution of tasks within the same old social bond, perhaps under the authority of a more efficient or more renewable Master: for *energy efficiency* and *renewables* to be a real discourse of transition and change, we need to look at how signification unfolds and whether they are attached to a chain of, for example, total energy volume, sufficiency, planetary boundaries. At the same time, current efforts at 'hystericizing' the discourse are not succeeding in transforming the discourse and are instead still resisted and repressed: nevertheless it is impossible to predict whether continuous 'hystericizing' acts aimed at questioning and challenging the dominant social link – perpetrated, for example, by a more progressive stance of a new Commission or a new Parliament or even new forms of radical activism – will eventually manage to cause an irreversible and painful rupture in the EU's climate mitigation discourse.

Conclusion

This chapter continued the investigation of the EU's climate mitigation action in the period 'in-between strategies' through the case of *energy efficiency* and *renewables*. The aim of the chapter was to challenge a seemingly closed discourse, expose its traumatic points, and ideally assess any possibility of disruption and transformation. Building on a previously debated understanding of knowledge (S2) as evidence-based disavowing its real command (S1), I zoomed in on the case of *energy efficiency* and *renewable*s because these policy tools play a key role within both the EU's medium-term and long-term strategies and because these are desirable policy tools which potentially carry a transformative potential. Via a Lacanian discourse analysis, I illustrated that overall *energy efficiency* and *renewables* represented the quintessence of mobilised, reductionist, bureaucratised, and valorised knowledge which took the form of legislative acts, technologies, targets, modelling within the climate and energy nexus. Yet by juxtaposing speech extracts as spontaneously produced, I emphasised the unintended effects of the signifying chain on the subjects observed and interviewed and I demonstrated that a Lacanian subject as an inconsistent identity emerged in the enunciation whenever *energy efficiency* and *renewables* are inconsistently signified. Thus, the key to understanding how discourse and subject are held together and presupposed lies in how the 'lack' emerging as gaps, inconsistencies, and blind spots, is handled in the discourse. More to the point, in the pursuit of achieving their climate mitigation goals, these split subjectivities as workers-consumers set 'knowledge' to work with the signifiers available to them, and this includes *energy efficiency* and *renewables*. Yet, rather than emerging as a disruptive force able to produce new significations, this ever-partial signification has been positively integrated into signification and turned into commodified knowledge and consumption objects. As a result, these subjectivities qua representatives of the discourse set the targets, draft impact assessments, exchange knowledge, build models, produce official policy documents, fund technology innovation and research, and enact smart, efficient or renewable consumption. This valorisation of the traumatic points of the discourse as a positive feature has often been accompanied by a paradoxical sense of (dis)satisfaction, that is, *jouissance* which fuels desire, as to its efficacy towards delivering climate change mitigation. As a result, the conservative reaction to the defiant elements of 'hystericization' – both deliberately placed during interviews and detected by the hysterical subjects – lead us to conclude that the EU's climate change mitigation action as formalised by *energy efficiency* and *renewables* is a formalisation of Lacan's University discourse – the discourse of commanded and disavowed knowledge – and of its parallel Capitalist discourse – the discourse of *jouissance* as commodified endless enjoyment.

Overall, *energy efficiency* or *renewables* as a presupposition of sense do not automatically deliver the desired, let alone required, reduction in GHG emissions. For the moment, within the 2030 Clean energy package and the 2050 long-term decarbonisation strategy – later Green Deal – we are facing a case of two signifiers that are well anchored by a very powerful dominant signifier from which they struggle to detach and are not able to shake the foundations of the discourse. In fact, every encounter with the 'lack', associated with a few powerless hysterics in the field, is resisted and their efficacy is ultimately hindered by the limits of the University and Capitalist social bond. As a result, the acclaimed energy transition qua climate action appears to be a fictitious change under the semblance of an efficient and perhaps renewable Master, supported by the fantasies of efficient products and renewable energy that promise to cover the impossibility of the discourse.

If this chapter has been concerned with signifiers that have a stabilised meaning in EU policy-making, what happens when a new signifier appears in the domain of the EU's climate action? The next chapter will introduce the case of the circular economy and will adopt the opposite perspective. Namely, if *energy efficiency* and *renewables* appear to have an apparent fixity of meaning that I have tried to open up and disrupt, the next chapter will investigate the journey of *circular economy*, from a floating signifier with a real rupture potential to its relative meaning stability.

Notes

1 The speaker refers in this case to the previous 20–20–20 package, setting a + 20 per cent target for renewables, not to the 2030 Clean energy package.
2 Translated from Italian (from a non-Italian native speaker).
3 Within the EU Parliament national parties are regrouped into European party groups. For example, the S&D Progressive Alliance of Socialists and Democrats regroups all the social democratic and socialist parties across the Member States.

6 The Lacanian case of the circular economy

A return of the repressed?

Introduction

If the previous discussion was concerned with signifiers that have a stabilised meaning in EU policy-making, what happens when a new signifier appears in the domain of the EU's climate action? This chapter continues the analysis of how the EU's climate action in the period in-between strategies constitutes itself through the subjects of the enunciation and introduces what was at that time an element of novelty in the EU's climate action, the circular economy. It will thus adopt the opposite perspective: if *energy efficiency* and *renewables* appeared to have an apparent fixity of meaning that I have opened and disrupted via a Lacanian discourse analysis, this chapter investigates if what was at that time a floating signifier with a rupture potential, *circular economy*, shook the foundation of an existing social bond or alternatively acquired relative meaning stability in a dominant social bond. This chapter and the next one will provide a 'symptomatic' reading in two parts – as we can detect two key phases – with the aim to assess its transformative potential.

Once again, starting from the signifying chain and focusing on its unintended effects on the subjects observed, I focus on how *circular economy* acquired relative meaning stability within the discourse. First, I argue that circular economy was initially made meaningful through a different battery of signifiers compared to the knowledge S2 that characterised the case of *energy efficiency* and *renewables* and provided insight into the potentially disruptive metaphor of 'circularity'. As a result, I illustrate how this signifier carried a potentially disruptive force that could have expanded the signification of rationalised and reductionist climate knowledge towards a holistic approach that considers the planet's interdependencies. At the same time, I demonstrate how an immediate reference to 'decoupling', that is, the idea of separating growth from environmental pressure, resulted in a breakdown of this alternative representation. For this purpose, I re-traced the signification journey of the circular economy within the EU and its subsequent linkage to climate action and showed how the circular economy is in principle recognised within the EU as an important part of effective climate action. In this context, a relevant aspect of the signification concerning *circular economy* is that it ideally leads us to consider the bigger picture of all material and organic flows and how these interact with the surrounding biophysical reality. Hence, a *circular economy* points to reorganising our

DOI: 10.4324/9781003378228-6

socio-economic relations and to the way knowledge is produced and exchanged, it implies greater symbiosis and coordination of all the EU policy DGs, a common pattern across policy areas, and would run against the reductionist and bureaucratised approach to knowledge production that has been so far distinctive of the Commission. However, my empirical investigation shows that for this reason it was strongly resisted to the point that the project was initially withdrawn precisely because of the difficulty in translating *circular economy* as 'measurement' knowledge at work, namely an impossibility to translating it into that quantifiable element that governs our societal and economic relations and directs policy-making. Finally, the chapter shows how, despite being met with resistance by both business stakeholders and some Commission DGs such as Clima and Energy, *circular economy* has eventually become part of the EU's policy-making in general, and climate action more specifically in the period 'in-between strategies'. At the same time, it seems to be integrated into a more reassuring and much less disruptive version which will be explored more in-depth in the next chapter.

The disruptive potential of circularity

The discussion in this chapter begins by providing insight into the potentially disruptive and 'hysteric' force that the metaphor of circularity brings. Circularity seems to be associated with a potentially alternative chain of 'metabolism', 'regenerative character', 'system complexity', and 'flux interrelation'. This alternative chain of signification seems to anchor *circular economy* to a linguistic network that considers all the interdependencies within the planet, as opposed to a simplified, rationalised, and reductionist knowledge that has been characterising so far the climate mitigation action. However, I will illustrate how an immediate reference to 'decoupling', that is to say, the possibility to separate economic growth from environmental pressure and resource extraction, resulted in the breakdown of this potentially alternative representation. As a result, in this chapter we are facing the case of a signifier at a signification crossroad: on the one hand, *circular economy* carries a disruptive force that attempts to produce an alternative signification with a wider ecological anchoring point, whereas on the other hand, it is dragged into the hegemonic signification machine driven by the Master Signifier of competitiveness, accumulation, and ultimate growth characterising both the University discourse of allegedly neutral knowledge and the Capitalist discourse of (false) enjoyment.

In Chapter 3, we learnt that Lacan does contemplate the possibility of transformation and that a change from one social bond to another is caused by a disruption that, as he puts it, 'hystericizes' the discourse (Lacan, 2007, 35). Nevertheless, he often warned his students about the force of the discourse, to the extent that what we might consider radical change might in fact retains the same logic of operation (Bracher, 1993, 73–74; Klepec, 2016, 116–117).

The term *circular economy* has not been coined by the EU but is a concept for which some proactive stakeholders such as the European Economic and Social Committee (EESC) or the European Environmental Bureau (EEB) lobbied through progressive DGs such as DG Environment. This aspect is relevant if we are to understand how this signifier is being pushed and pulled across different signifying networks by a multitude of different subjectivities, and how it is being 'imported' and spoken of within the EU. When speaking of the circular economy today, stakeholders tend to refer to the version provided by the Ellen MacArthur Foundation. For example, an EESC speaker, one of the stakeholders actively promoting the circular economy (see Table A.2 in Appendix), referred to it in the following way:

> I think that, ok, the philosophy of the circular economy that I adhere to is the same as the one presented, *mostly the one presented by the Ellen MacArthur Foundation.* Even though I don't agree with them on every approach but the philosophy of circular economy, *which is very close to the original*, to Walter Stahel. And that is in simple terms that there's two, ok there's many, many circles. But there' s two main circles.
>
> (Interviewee 5, 30 October 2018)

The above extract informs us about the existence of something that we might refer to as the 'original philosophy' of the circular economy and that there is some convergence in terms of adhering to the version presented by the Ellen MacArthur Foundation. Therefore, at this preliminary stage of discussion, it is worth investigating what the Ellen MacArthur Foundation says about the circular economy. On their website, we learn that the 'circle' or 'cycle' metaphor is an ancient one, but this idea of circularity experienced a revival in the post-war period in the attempt to grasp the complexity and interrelation of some systems:

> The notion of circularity has *deep historical and philosophical origins.* The idea of feedback, of cycles in real-world systems, is ancient and has echoes in various schools of philosophy. It enjoyed a revival in industrialised countries after World War II when the advent of computer-based studies of *non-linear systems unambiguously revealed the complex, interrelated, and therefore unpredictable nature* of the world we live in – *more akin to a metabolism than a machine.*
>
> (Ellen MacArthur Foundation, 2019)

Thus, the 'circle' metaphor reminds us of the complex interrelations that keep systems together, for example, our ecosystem(s). More specifically, this idea of 'circularity' seems to carry an innovative force from the point of view of signification insofar as it is attached to a chain of 'complexity', 'interrelation', and 'metabolism'. This chain seems to be antithetical to the 'line' metaphor that has characterised our way of exploiting natural resources and organising our economy, namely along the line produce-use-dispose, with consequent GHG emissions related to those processes. Furthermore, by opposing 'metabolism' and 'machine', it might be argued that an idea of circularity stands in opposition to those ways of knowing that fail to grasp this system's complexity and that have been previously referred to as rational and reductionist, often reinforced by a rigid bureaucratic organisation of work. A further component of this metonymy of circularity is its regenerative character:

> *regenerative by design* and aims to keep products, components, and materials at their highest utility and value at all times, distinguishing between *technical and biological cycles.*
>
> (Ellen MacArthur Foundation, 2019)

The regenerative component of a circular economy aims to save natural resources and raw materials and reuse what is already in circulation, and it is thus associated with resource scarcity. As the extract explains, we must first distinguish between two main circles: the natural biological organic flows such as food, and the material flow, that is all the non-natural manufactured products that populate our everyday life and that are the focus of this chapter.

Moreover, the EESC speaker below explains how this signifier floats across the linguistic network by introducing a new series of concepts such as eco-design, remanufacturing, repairing, planned obsolescence,[1] recycling, service-based model which will return at a later stage in the analysis:

> So, the concept of circular economy and how to address that clearly is that first of all there are 2 big circles. One is to discuss *material flows*, hard raw materials and how they move within an economy. That's where *eco-designing products comes in*, where you have a *remanufacturing, repair, and the planned obsolescence is eliminated* and products can be reused or can be modular. So, there are components, parts that can be reused. That's the material flow basically and that's where recycling, the service-based model as new business model, as new ownership model, fits into that. But then there is *the biological or ecological flow* of materials which is organic natural materials.
>
> (Interviewee 5, 30 October 2018)

This speaker introduced some new relevant signifiers in the chain of signification that will return later in the analysis. At this stage, it is enough to highlight how the notion of circular economy can be explained by reference to 'eco-design', which introduces concepts such as 'repairing', 'remanufacturing', and 'eliminating planned obsolescence'. These in turn slide in a chain of 'recycling', and of 'service-based business models' which translates the abstract understanding of circular economy into practice.

Although the speaker above mentioned that there is a version 'which is very close to the original, to Walter Stahel' (Interviewee 5, 30 October 2018), the concept of the circular economy is not as straightforward as it seems for two main reasons. First, today the notion of circular economy appears to be spread across different schools of thought and cannot be traced back to one scholar or philosophy only. Thus, rather than speaking of a singular circular economy, we can speak of different 'circular economies' tied together by a metaphor of loops and circularity.[2] Second, the binary distinction of the material versus the organic is not as clear-cut, which contributes to blurring the conversation as our EESC actor explains (sic):

> And I don't mean bioplastics by the way because bioplastics, well because organic flow of material is usually food or trees or materials that are made of 100 per cent naturally produced and they go back in through biodegradation back into the ecological and natural system. Bioplastics are drawn, in some cases, *from natural biological world and manipulated* to come to the material flows, and because they are no longer biodegradable in the technical sense of the word, they essentially biodegrade down to their constituent parts. *But sometimes they're manipulated so much that it takes specific conditions for them to biodegrade.* A lot of the compostable cups, they end up in the landfill because they can't actually go in a composter. They don't break down, so even though they're biological materials, or bioplastics or bio packaging. *They leave the biological circle and they become part of the material circle.* And you need to manage with that in mind. *And I think the concept gets blurred* in the conversation. But also, this is one of the comments of the overall philosophy of it if that makes sense.
>
> (Interviewee 5, 30 October 2018)

From the above extract, we learn that the material versus organic distinction is not as neat as it seems because we can have biological materials that are manipulated and enter the material circle, such as bioplastics. This is what, in the speaker's view, contributes to blurring the conversation of the overall philosophy, but as we will see in the next section this is not the only fuzzy aspect. What it is important to highlight at this stage is the potential alternative signification that a metaphor of circularity can bring in terms of metabolism, regeneration, and complexity, as opposed to a linearisation, reductionist view of the organic and material fluxes within an economy. In other words, *circular economy* at this stage can be 'that which' 'hystericizes' the discourse (Lacan, 2007, 35), that element of disturbance into a hegemonic system of signification. At the same time, in the next section, we will see how a challenging element in the conversation results in a potential breakdown of representation of this 'circularity' metaphor. In fact, the reference that the Ellen MacArthur Foundation makes to 'decoupling'[3] casts doubt over the direction that the circular economy is taking and how it is being promoted:

> This new economic model seeks to ultimately *decouple global economic development from finite resource consumption.*
>
> (Ellen MacArthur Foundation, 2019)

Circular economy was being attached to a chain of 'complexity', 'interrelation', and 'metabolism', but the above extract interrupts a potentially alternative signification and results in a potential breakdown of representation. It seems that *circular economy* is sliding back into the signification of 'decoupling', that is, the idea of separating economic growth from environmental degradation. As a result, it seems that we are facing an interesting case. On the one hand, *circular economy* is a potential hysterical and disruptive force that attempts to produce an alternative signification with a wider ecological anchoring point. On the other hand, available social-discursive representations regulate its meaning and provide it with a stable – yet ambiguous and fragile – identity by dragging in the signification machine that regulates the University discourse of knowledge and the Capitalist discourse of enjoyment. As we will see in the next section, today *circular economy* is recognised within the EU as an important part of an effective climate action, which somehow indicates the desire of achieving the required emissions reductions.

Circular economy and climate action: desire in motion

The previous section has highlighted the hysterical character of 'circularity' and pointed to the potentially alternative signification that 'circularity' might bring and argued that the circular economy signification ideally forces us to bear in mind the bigger picture of all material and organic flows rather than work on small narrow signification spaces, as emphasised in the case of the climate-energy nexus. Yet, it has also highlighted how an immediate reference to the more stable discursive representation of 'decoupling' resulted in a breakdown of this alternative signification.

Today *circular economy* is in principle recognised within the EU as an important part of an effective climate action, even in the recent European Green Deal, which might be indicative of the subject's desire to cut achieve the ultimate emissions reductions. In this respect, the necessary and desirable role of the circular economy is recognised by DG Clima, which is not a lead DG in the matter of circular economy:

> The response is simple, if you want to solve the climate problem *there is no way that you can go without the circular economy* because we will have 30 per cent more population by 2050.
>
> (Interviewee 10, 21 November 2018)

Similarly, the following DG Grow speaker explains the 'common-sensical' and 'evidence-based' character of the circular economy:

> Well, I'm aware of the principle of the circular economy . . . , we have colleagues here working full time on the principles of circular economy. *It's one of these policies that seems common sense, it's like evidence-based policy-making. I mean since when policymaking hasn't been evidence-based.* It's providing an aim for something which has always been there as a reflection on how you make things. Why would you set out to make something which couldn't be recycled or reusable, in fact you may not have a choice over there, it's not a deliberate policy end.
>
> (Interviewee 11, 22 November 2018)

In these extracts, the *circular economy* is spoken of as a necessary component of climate action: 'There is no way you can go without the circular economy' (Interviewee 10, 21 November 2018). Its decisive role is almost presumed and taken for granted as 'one of these policies that seems common sense' (Interviewee 11, 22 November 2018). It seems that at least on the side of the EU representatives there is a desire to come up with an effective climate action and give space to those contributions that help pursue this objective. Furthermore, as anticipated, *circular economy* ideally forces to bear in mind the bigger picture of material and organic flows rather than work on small narrow signification spaces, which are formalised, for example, in the Commission's division of work. Hence, it can be beneficial to expand the signification horizons of climate action as energy transition detected in the previous two chapters, in that it ideally requires consideration of flows and how these interact with the surrounding biophysical environment. In this regard, the *circular economy* could imply greater symbiosis and coordination of all EU policy DGs, a common pattern across policy areas, and would run against their reductionist and bureaucratised approach to knowledge production that has been so far distinctive of the Commission.

The willingness to overcome the 'silos' approach emerged at two events I participated in, which seem to suggest an intention to work with a more integrated approach. The first event was called 'Tackling Planned obsolescence' (30 November 2018) (see Table A.2 in Appendix).[4] On this occasion, it was reproached that despite the almost unanimous adoption of the EESC Opinion (2013) and the European Parliament resolution of 2017, the Commission has been slow to respond and lacked a holistic approach, or rather lacked a willingness to adopt a transversal approach for all political sectors. It has been reported that the (at that time) First Vice-President of the EU Commission, Frans Timmermans, and Vice President, Jury Katainen, had expressed their commitment to deal with the planned obsolescence in reaction to the European Parliament rapport showing some progress, but this progress has been very slow. However, Marie Paule Benassi from DG Just, Acting Director[5] shortly claimed that the current Commission was no longer working in silos:

> Since the Juncker Commission *we have stopped to think in silos,* with all the services of the Commission adopting a coordinated approach.
> (Marie Paule Benassi, 30 November 2018)

Similarly, during another event called 'Boosting circularity in SMEs', that is in Small and Medium Enterprises (6–7 December 2018, see Table A.2 in Appendix), the keynote speech of the Commission represented by Peter Czaga, Policy officer from DG Environment, pointed to the need to push the circular economy to all policy areas:

> *The need to streamline circularity, not only in environmental policy but in all other policy areas*, enterprise policy, industrial policy, finance, education and so forth. Katainen and Timmermans said *we need to streamline the circular economy in all policy areas.*
> (Peter Czaga, 06 December 2018)

As a result, the role of the circular economy is advocated as desirable, evidence-based, and commonsensical and would imply greater integration and symbiosis between DGs within the Commission, which could potentially translate into a new type of knowledge (S2).

In this respect, it is worth investigating how *circular economy* knowledge is put to work, while observing the effects of this signifier on the people interviewed and observed. Although at present the role of the circular economy is advocated as a necessary component of an effective climate action, this has not always been the case. In fact, the *circular economy* within the EU can be said to have been perceived at the very beginning as a disruptive element in representation. This would seem to be confirmed by the simple observation that the first phase that characterised the entry of the circular economy within the policy-making process of the EU, in general, is that of 'resistance'.

Circular economy: the phase of resistance

In this section, we will observe that despite being recognised as an important part of the EU's climate action and EU policy-making in general, the *circular economy* within the EU has been perceived at first as a destabilising element in representation, to the point that it was resisted for not conforming to the status quo system of knowledge. This element of disturbance points first to the fact that an idea of 'circularity' in opposition to the 'line' might have been ostracised by business actors because it forces us to re-think the economy as a cycle, which can, in turn, introduce an idea of limit, pace, and size rather than business at all costs at any time. Second, this element points to the way 'knowledge' is produced and exchanged within the EU.

Interviews in the field revealed that introducing a *circular economy* into the EU's decision-making procedure has not been a smooth process. Rather, it has been met with evident resistance which resulted in the withdrawal of the project. In this respect, during an interview with DG Environment that took place shortly after the publication of the long-term strategy, the *circular economy* came up as a free association in the conversation on citizens and their concerns over climate change, environment degradation, air pollution, and planet degradation in the wider sense. More specifically, the EU representative spontaneously spoke about the element of resistance that surrounded the circular economy throughout its journey within the EU:

> What is the political and bureaucratic history of the circular economy? There was a Communication that was presented shortly before the Juncker Commission started its term. *And one of the first things that this Commission has done is to withdraw the circular economy. But it was rumoured that there was a big pressure from European companies, big companies.* But how can you withdraw this if for us this is a priority? We are already doing the circular economy, we're talking about energy transition, but there is also a circular economy transition that is also present, this is very clear. Brussels is

recycling I got the figures 90 per cent of paper, 80 per cent of plastics. There is much research that shows how the circular economy is a sine qua non for decarbonization. . . . *After a lobbying activity, the Commission position was, we have withdrawn that to make it better and make a more ambitious one,* which is the one that we got today, that it's the one that is becoming more and more important within the Commission. . . . In the strategy that has been presented yesterday (the long-term strategy, ed.), it is very present.

(Interviewee 14, 29 November 2018)

From this account, we understand that the first proposal on the circular economy was withdrawn as soon as the Juncker Commission stepped in, following a lobbying activity by business actors. The same story came up with more details, during the interview with an EESC member:

The history was the previous Commissioner of the Environment Jannus Petocnik, he pushed for the Circular economy package. There was a lot of initial study by a guy called Professor Walter Stahel . . . he's known as the grandfather of the circular economy . . . and since the 1970s he's been writing about circular economy. . . . Walter had been writing about this as a concept where commissioner Petocnik got into office *he said OK let's do this and he published a very ambitious Circular economy package, and this was during the recession you know in the global downturn.* It was exciting because he proposed a new economic model. *And it was published, and it started going through this process here, in the parliament and then due to a very strong lobbying from different groups in particular Business Europe it was removed, it was pulled.* And at that time there was lot of upset and anger that it was removed from the agenda completely and the anger was building and building until there was some strong protesting particularly in the Brussels bubble from the EEB. You can imagine the European Environmental Bureau and at some point, in 2015 Commissioner first Vice President Timmermans said at some conference, he said, circular economy – why did you remove this form the agenda? And he said the only reason it was removed was to make it better, to make it more ambitious and to make it even greater and everybody went really? And the feeling was that that was a signal that perhaps didn't come from Juncker. . . . We're going to produce something even better, and in December 2015, . . . while everyone is in Paris and focused on this celebrating and excitement, here quietly the Commission produces here's our proposition, a new improved more ambitious proposition. And what got lost in the noise of Paris was this less ambition and lower targets. It was less ambitious in one analysis as it represented less of carbon reductions than the withdrawn package. *So, on the one hand the Commissioner said we're going to produce something even better. But what was produced, the analysis, showed something much less percentage targets, that lower the ambition.*

(Interviewee 5, 30 October 2018)

Although the two versions from the EESC stakeholder and the Commission differ in the level of ambition between the two proposals, these two accounts converge on the circular economy being met with resistance. Both accounts agree on the fact that the package had been first withdrawn and then brought back in by an opposite strong pressure, for example, by the EEB. Although it is impossible to establish the exact cause of such resistance by business actors,[6] we can question whether introducing a *circular economy* has been perceived as a potentially powerful element capable of challenging the dominant social link and the knowledge underpinning this social link. We might guess whether an idea of 'circularity' in opposition to the 'line' might have been ostracised by business actors in that it forces them to re-think the 'business as usual' logic and to re-think the economy as a cycle. This could introduce an idea of limit, pace, and size rather than business at all costs at any time.

Interestingly, on the side of the EU Commission, the disruptive element that caused resistance points to the way 'knowledge' is produced and exchanged within the EU. In the previous chapters, I provided examples of a type of knowledge that is rational, quantifiable, and measurable and argued that what does not fit the measurement paradigm is automatically discarded, the type of knowledge that has been the subject of Lacan's critique in his theory of the four discourses. In this case, the 'resistance' towards *circular economy* within the Commission points precisely to the difficulty of the circular economy to be modelled, and thus to the impossibility to count as climate action knowledge, as DG Environment explains:

> And then at the beginning of the circular economy, these colleagues in (DG, ed.) Clima and Energy didn't see it as . . . . *ah but we cannot model this*, he (a member of the EPSC,[7] ed.) insisted and had an influence to reach an agreement with Grow (DG, ed.) and pressure them.
>
> (Interviewee 14, 29 November 2018)[8]

The same EU representative from DG Environment continued to speak of the problem of modelling that caused this resistance from DG Clima and DG Energy in the early debate of the preparation of the current long-term strategy. Eventually, the circular economy is not modelled but assumed as a parameter,[9] which, as it is, does not refer to how much circularity can be in fact put into practice:

> *But I can say this: there has been a debate* when the process of preparation of this long-term strategy began. *One of the problems was modelling.* In all the modelling on climate change there is no circular economy. *And they came with assumptions of not modelling the circular economy, but introduce circular economy parameters on the strategy, in the modelling.* So, if you look at this strategy there are 8 scenarios, some technological, one is circular economy, of these scenarios, you must check the data, if I remember well, one is hydrogenated synthetic fuels, another is circular economy, another is resource efficiency, energy efficiency etc. with 5 we would get to 80 per cent reductions.
>
> (Interviewee 14, 29 November 2018)

Several aspects emerged in this conversation which in turn build on previous discussions regarding the EU's climate mitigation knowledge. First, what emerges is the discussed pattern of quantification, rationalisation, and valorisation that characterises the allegedly neutral discourse in which climate action is defined, the hegemonic modern science of the University discourse. According to this knowledge paradigm, each concept is spoken in terms of measurement, which in turn defines the EU as a technical, 'regulatory' subjectivity. In the above extract, the resistance towards *circular economy* qua modelling practice is resisted precisely by those DGs that deal with climate action modelling, rather than by those who try to keep the climate signification space open, as DG Environment. Thus, the element of resistance can perhaps be explained by the non-conforming character of circular economy with the hegemonic knowledge, that is, the difficulty in translating *circular economy* as 'measurement' knowledge at work, and an impossibility to translate it into that quantifiable element that governs our societal and economic relations and directs policy-making.

In summary, the journey of the circular economy started with a signifier that was initially perceived as a challenging element in signification, something that could potentially undermine the anchoring point of representation or – to put it in Lacanian terms – hystericize the discourse by opposing a holistic approach of ecosystem interdependencies to a view of linear flows and related quantified reductionist solutions to climate change. The actual withdrawal of the project, as soon as Juncker's Commission started its term, can be said to have shaken the foundation of the discourse by implying rethinking our societal and economic relations and how knowledge is produced and exchanged, the circular economy project was at first resisted. However, despite being met with resistance by both business stakeholders and some Commission DGs such as Clima and Energy, in the next section, we will see that *circular economy* has eventually become part of the EU's policy-making in general and then climate action more specifically, thanks to the mobilisation of some stakeholders such as EESC and some progressive DGs such as DG Environment. However, if *circular economy* initially carried the disruptive element of 'circularity' in opposition to the 'line' – to retain the original metaphor – it now seems to be accepted within the dominant social bond.

Circular economy and climate action: the phase of acceptance

Today, the metonymic link 'circular economy' + 'climate action' within the EU is something achieved and taken for granted and the EU is currently working within the framework of a new, updated (2020) Circular Economy Action Plan which is also one of the building blocks of the European Green Deal (DG Environment, 2024). Yet, this has not always been the case, because when it was first introduced it was mainly associated with resource scarcity, regardless of any emissions mitigation effects.[10] At the same time, I have illustrated that the current climate change mitigation action within the EU can be said to be a formalisation of Lacan's University discourse of allegedly neutral knowledge and the Capitalist discourse of false enjoyment. With reference to the period 'in-between strategies', we might wonder how *circular economy*, originally a hystericizing element in the dominant

University and Capitalist discourses to the extent that it was resisted, is now embedded in this new 'acceptance' phase.

When the *circular economy* was first introduced within the EU and before the first proposal was withdrawn, it was associated with finite resources and resource scarcity, thus the environment and the planet in its wider understanding, but not specifically with 'climate action'. However, it can be argued that the link with climate policy refers in this case to an association with existing policies that treat climate change within the climate + energy signification space, rather than within an alternative chain climate + environment. In this respect, DG Environment explains:

> I suggest you look at the *Communication on the circular economy because it didn't speak much of climate change, 2–3 times. This has consequences on change* because probably there was no strong scientific evidence to demonstrate this impact with data. Now there is.
>
> (Interviewee 14, 29 November 2018)

Indeed, if we look at the first EU Commission Communication 'Towards a European Circular Economy' COM (2014) 398 final/2, the link between the circular economy and *climate action*, is not that prominent:

> *Some EU policies* and instruments already *provide tools and incentives in line with the circular economy model. The waste hierarchy* that underlies our waste legislation is leading progressively to adoption of the preferred options of waste prevention, preparation for reuse and recycling, and discourages landfilling. *Chemicals policy* aims at phasing out toxic substances of very high concern. *Some eco-design measures for energy-related products* include requirements on durability and to facilitate recycling. The Bioeconomy Strategy promotes the sustainable and integrated use of biological resources and waste streams for the production of food, energy, and bio-based products. *Climate policy creates incentives to save energy* and reduce greenhouse gas emissions.
>
> (COM (2014) 398 final/2, 5)

The second Communication on the circular economy is called 'Closing the loop – An EU action plan for the Circular Economy' (COM (2015) 614 final, 2). It was released the following year, in 2015, and is the foundation of the first (2015) Circular Economy Action Plan, now updated by the 2020 Circular Economy Action Plan (CEAP, COM/2020/98 final). Moreover, this second Communication addressed the link between the circular economy and climate policy explicitly in relation to climate change mitigation efforts:

> The circular economy will boost the EU's competitiveness by protecting businesses against scarcity of resources and volatile prices, helping to create new business opportunities and innovative, more efficient ways of producing and consuming. . . . At the same time, it will save energy and help avoid the

irreversible damages caused by using up resources at a rate that exceeds the Earth's capacity to renew them in terms of climate and biodiversity, air, soil, and water pollution. *A recent report also points at the wider benefits of the circular economy, including in lowering current carbon dioxide emissions levels.*

(COM (2015) 614 final p. 2)[11]

These extracts from two different Communications – before the circular economy project's withdrawal and after its reintroduction, respectively – are relevant for one main reason: it seems that *circular economy* has been linked since its introduction to all the existing policies (and signification spaces) that deal with the environment and the planet in its wider understanding, for example, the existing legislations on waste, chemicals, eco-design measures, but not specifically to 'climate action'. Hence, the metonymic link with climate policy is perhaps spoken of by the DG Environment speaker as an association with existing policies that treat climate change within the climate + energy signification space. Apparently, an alternative chain climate + environment that treats climate change within a signification chain of planet, ecosystems, and biophysical environment at large is excluded. Thus, the DG Environment speaker seemed to suggest that the link between *circular economy* and climate policy refers to a specific meaning of 'climate action', namely that of the climate change qua energy transition metonymy, the embodiment of Lacan's University and Capitalist discourses previously discussed in this book.

Unlike the case of *energy efficiency* and *renewables* in which an element of continuity with climate action could be immediately established between the 2030 Clean energy package and the 2050 long-term strategy, the case of *circular economy* and its link to climate action is not as straightforward and pertains to a separate policy document, the first Circular Economy Action Plan (2015), which had been debated for 2 years (Interviewee 5, 30 October 2018). Notably, the relevance of *circular economy* for climate action in the period in-between strategies is strengthened by the prominent role accorded to it in the Communication on the 'long-term strategy' (COM (2018) 773 final). This newly found prominent role is further exemplified by the fact that DG Clima became one of the public faces of the circular economy while previously this role was exclusively played by DG Environment.

For example, during the COP24, I observed a side event at the EU pavilion called 'Circular Economy, the missing link in climate action?',[12] to establish a link between *climate action* and *circular economy*. Yvon Slingenberg, Director of DG Clima[13] and one of DG Clima's main public representatives, gave a keynote speech (11 December 2018) and established a clear connection between *circular economy* and climate action. Within this framework, the role of the Commission as a knowledge exchange facilitator is re-affirmed:

Circularity has a key role to play. We, from the EU Commission perspective, we are *administrators, regulators, supporting the policymakers and we put in place rules, statistics framework, regulatory framework* in order to support and help business in the right direction transition and also, of course, *maintain the competitiveness of EU business, but in the clear direction of a more sustainable economy.* We are convinced that Europe can lead in this transition . . . I think it's on the map the EU Circular economy action plan which has also made a clear case for reducing GHG reductions from circularity that was put forward in 2015. In the meantime, you also mentioned the International Resource Panel has already come up with a different figure showing how significant the GHG reductions from circularity can be. I have impressive figures here 63 per cent, at the same time boosting economic activity in the EU and increased GDP by 1.5 per cent.[14] *So clearly, we win and that's the message we're trying to bring.*

(Yvon Slingenberg, 11 December 2018)

This metonymic link between *circular economy* and *climate action*, purported by technical and regulatory subjects with quantitative data at hand, seems to be perfectly in line with Lacan's University and Capitalist social bonds of *energy efficiency* and *renewables* previously discussed. For example, a section of the official long-term strategy policy document 'A competitive EU industry and the circular economy as a key enabler to reduce greenhouse gas emissions' states:

> *The EU industry is already today one of the most efficient* globally and this is expected to continue. *A competitive resource-efficient* and circular economy will need to develop to keep it so. The production of many industrial goods like glass, steel and plastics *will see further significant reductions in energy needs and process emissions, particularly with increasing recycling rates.* Raw materials are indispensable enablers for carbon-neutral solutions in all sectors of the economy. Recovery and recycling of raw materials will be of particular importance in those sectors and technologies where new dependencies might emerge, such as a reliance on critical materials like cobalt, rare earths or graphite, whose production is concentrated in a few countries outside Europe.

(COM (2018) 773 final 12)

The official policy document reaffirmed once again the importance of *circular economy* in the function of the competitiveness of the EU industry. Unsurprisingly, in the document, we find again all the common themes that characterised the climate mitigation qua energy transition reflections in the previous chapter. For example, we find reference to the role of the circular economy in bringing 'significant reductions in energy needs and process emissions', thus in terms of 'savings' and 'efficiency', but most certainly not a reduction in volumes of production:

> *Given the scale of fast-growing material demand, primary raw materials will continue to provide a large part of the demand.* But *a reduction of materials input* through re-use and recycling will improve competitiveness, create

business opportunities and jobs, and require less energy, in turn reducing pollution and greenhouse gas emissions.

(COM (2018) 773 final, 12)

Finally, we identify once again the recurring theme of the subject-consumer governed by the untouchable command (consume and enjoy!) which is invited to divert consumption patterns towards allegedly 'circular' solutions. These can be either induced by technological development, that is, digitalisation, or are promoted as allegedly 'active' consumption choices of eco-friendly products and services:

> *Consumer choices will also matter for product demand. Some may come from other ongoing transformations*, such as digitalisation reducing paper demand. *Others will be more climate conscious choices, such as customers increasingly asking for climate and environmentally friendly products and services.* This requires more transparent information to consumers about carbon and environmental footprints of products and services so that they can make informed choices.
>
> (COM (2018) 773 final, 12)

In this respect, a speaker from DG Environment makes a spontaneous but very relevant point about the wording 'consumer choice'. When describing the different scenarios modelled in the long-term strategy the speaker explained:

> Then they make a combination technological plus circular economy, and then another combination circular economy and lifestyle changes. They do not use this term; however, *they use consumer choice. At the beginning it was lifestyle, but it seems that people in the Berlaymont*[15] *did not like it because they feel it's like imposing a way of living.*
>
> (Interviewee, 14 29 November 2018)

The discussion about the fear of 'imposing a way of living' which took place at the high political levels of the Commission is a recurrent theme in the speech and links back to the previous discussion on how the alleged neutral knowledge (S2) works as a presupposition shaping our free-thinking, analyses, and apparently unbiased scientific inquiries but in fact disguises its performative and fictional character under a flat and apparently objective knowledge which in truth works for the Master Signifier (S1) (Pavon Cuellar et al., 2010, 264–265). In this case, the debate between the wording 'lifestyle change' and 'consumer choice' is significant as lifestyle change introduces a stronger excess of meaning 'change' compared to consumer choice. The latter keeps the alleged 'freedom to choose' unaltered, or rather our freedom to choose what to consume, with consumption being treated as a factual thing that cannot be questioned. However, more precision is needed here: the speaker is right in stating that in the 25-page policy document, that is, the Commission Communication (COM (2018) 773 final), *lifestyle change* is replaced with *consumer choice*. This is perhaps due to the fact that Communication is a document that has a wider echo and will be read by a wider range of actors in society. In the supporting 400-page staff working document, the term *lifestyle change* is retained and is also one of the predicted scenarios (COM (2018) 773 final supporting analysis). In any case, by juxtaposing or substituting *lifestyle changes* with *consumer choices*, and establishing thus a homologous relation between the two, it is possible to infer that the real change in 'lifestyle' is applicable only to the individual buyers-consumers. This then would not mean a paradigm change at the collective level including modes of production on a large scale, and volumes of production across policy areas, for instance, in industry, transport, or agriculture.

As a result, after an initial phase of 'resistance', the *circular economy* has been reintroduced and accepted, most notably as an ally to climate action – or rather a specific understanding of it. At the same time, it seems to be co-opted in a more reassuring version within the hegemonic social bond. In the examples above, we can appreciate that all signification produced has slid into the hegemonic discourse and is made meaningful and stable by the same command of competitiveness and growth that gives direction to the overall discourse. Yet, this is not the end of the story. Dismissing the case of *circular economy* as an easy case of co-optation into the dominant University discourse of disavowed knowledge and/or Capitalist social bond of *jouissance* is too simplistic and precludes a whole series of insightful reflections about its transformative potential. Indeed, as we have seen at the beginning of this chapter, the *circular economy* did not originate in the function of the disavowed dominant signifier of consumption, competitiveness, and growth. Rather, unlike *energy efficiency*, which can be regarded as the by-product of the University discourse of 'knowledge at work', the *circular economy* was a case of a 'hysteric' knowledge which originally drew on an alternative philosophy and metaphor of complexity and metabolism that has only been imported into the dominant University and/or Capitalist social link at a later stage. At the same time, a seemingly closed discourse is in fact to be considered open due to the presence of the inconsistent Lacanian subject of the enunciation, produced by language as a surplus of sense, who is driven by the desire to reach full representation within their socio-symbolic structure. As a result, we can still observe the effects that *circular economy* has on the subjects observed and interviewed – that is to say, we can investigate what 'becoming circular' means for the EU – to detect if any fractures in the discourse are still produced and reflect on how these fractures are handled. We might wonder, for instance, whether the more recent metonymy circular economy + climate action has implications for how each DG sees circularity and how their desire to come up with a comprehensive climate policy mobilises their knowledge in this respect. More to the point, we might wonder whether it is spoken of as a new centre around which our socioeconomic relationships must be re-thought and a common pattern across policy areas or whether each DG associates those circular aspects linked to their specific tasks. Therefore, the next chapter investigates how *circular economy* is produced as an effect on the enunciation in the 'acceptance' phase with the aim to assess whether any traces of its disruptive potential remain and if and how we can still differentiate it from other 'green growth' concepts such as energy efficiency.

Conclusion

This chapter has continued the investigation into the EU's climate mitigation action within the period in-between strategies and introduced the case of the circular economy to investigate the potential of shaking the foundation of the dominant University and Capitalist discourse. This exercise has not been a critique of the circular economy per se; rather, the critique is addressed to how 'circularity' is spoken by its actors in relation to delivering an ambitious climate mitigation programme. The chapter started by providing insight into the disruptive force that the metaphor of 'circularity' carried with it and argued that it appeared as a potential hystericizing element challenging the University discourse as mere rationalisation, insofar as it could have expanded signification towards a holistic approach that considers issues of metabolism, regeneration, complexity, and interrelation. However, an immediate reference to 'decoupling' resulted in the breakdown of this alternative representation and consequently, I re-traced the journey of the circular economy within the EU starting again from an analysis of the signifier, the unfolding signifying chain, and its effect on the speaking subjects. More to the point, I illustrated that *circular economy* is in principle recognised by the speakers as an important part of effective climate action and as such it constitutes another discursive representation in the battery of signifiers which attempts to resolve the 'lack' of an ever-failing discourse. I also argued that ideally, *circular economy* could re-imagine policy knowledge and practice as it would imply greater symbiosis and coordination of all EU policy DGs, a common pattern across policy areas, and would run against their reductionist and bureaucratised approach to knowledge production that has been so far distinctive of the Commission. This is because it ideally requires a consideration of flows and how these interact with the surrounding biophysical environment. At the same time, I demonstrated how the initial perceived element of disturbance of 'circularity' as opposed to the 'line' has been gradually resisted through an actual withdrawal of the project because it would have necessitated discussing our societal and economic relationships are organised and how (and what type of) knowledge is produced and exchanged. In fact, its re-introduction in the 'acceptance' phase as well as its new association with climate action – which in principle could regarded as a positive achievement – pointed to a co-optation into the hegemonic climate action social bonds of disavowed knowledge and *jouissance* in a less traumatic and more reassuring version. Yet, as the discourse is always fractured by virtue of the Lacanian subject of the enunciation, in the next chapter, I will observe the effects of the *circular economy* on the subjects willing to bring effective climate action to detect whether discursive ruptures still emerge and if and how they can be leveraged upon.

Notes

1 Products that are designed to stop working more rapidly, usually shortly after the expiry of their guarantee (EESC, 2019).
2 These are the performance economy (functional service economy) by Walter Stahel; the cradle-to-cradle design philosophy by William McDonough and Michael Braungart; the

biomimicry by Janine Benyus; the industrial ecology by Reid Lifset and Thomas Gradael; the Natural capitalism by Amory and Hunter Ovins and Paul Hawken; finally, the blue economy systems approach by Gunter Pauli. It is beyond the purposes of this work to go into the depth of each single school of thought, but it is sufficient to acknowledge that this fragmented categorisation is contested in the literature (see Korhonen et al., 2018).

3 This concept has gained importance in the UNSDGs agenda and now it is a common term accepted in the sustainable development semantic field. SDG 12 'Ensure sustainable consumption and production patterns' cites: 'One of the greatest global challenges is to integrate environmental sustainability with economic growth and welfare by decoupling environmental degradation from economic growth and doing more with less. Resource decoupling and impact decoupling are needed to promote sustainable consumption and production patterns and to make the transition towards a greener and more socially inclusive global economy' (UNEP, 2024).

4 EESC event to address planned obsolescence to protect consumers and transition to a circular economy.

5 Directorate-General for Justice and Consumers.

6 Business Europe declined my request to be interviewed.

7 The European Political Strategy Centre (EPSC) is the European Commission's in-house think tank, established in November 2014 by Former President Jean-Claude Juncker. This think tank operates directly under his authority (EPSC, 2019).

8 Translated from Italian from a non-Italian native speaker.

9 The discussion within the supporting analysis around the limitation of modelling recognizes the limits of working with assumptions.

10 The European Commission launched its first Circular Economy Action Plan in 2015 and this is the one mentioned in this project. On 11 March 2020, the Commission launched a new Communication on a new Circular Economy Action Plan (COM (2020) 98 final).

11 This refers to the report *Growth within: a circular economy vision for a competitive Europe*, report by the Ellen MacArthur Foundation, the McKinsey Centre for Business and Environment and the Stiftungsfonds für Umweltökonomie und Nachhaltigkeit (SUN), June 2015.

12 Although in those days the EU was running a series of official long-term strategy side events, this specific event was not an official event, but one of the many side events that are organised by the Commission.

13 Responsible for International climate negotiations and mainstreaming of climate issues in EU policies.

14 The reference is to a study that is also mentioned in the long-term strategy supporting analysis on page 144, Material Economics AB (2018), The Circular Economy.

15 This is the name of the building in the Commission where the actual political decisions are taken.

7 The Lacanian case of the (non) circular economy

Introduction

In this chapter, I continue the story of the circular economy and discursively follow its reintroduction within the EU's climate action discourse in the 'acceptance' phase. The previous chapter suggested that by its reintroduction, the circular economy might have lost the disruptive and potentially liberating potential that characterised the original idea of 'circularity' to become co-opted in the same University discourse of disavowed knowledge and the Capitalist discourse of *jouissance*, which characterise energy efficiency and renewables and in which ruptures are endlessly commodified. At the same time, I have previously demonstrated that a seemingly closed discourse is in fact shown to be open due to the presence of the subject of the enunciation driven by the desire to reach full linguistic representation, that is, in this case to achieve an ideal and effective climate action. Therefore, it is still possible to investigate what the reintroduction of *circular economy* within the EU has entailed and observe the effects of *circular economy* on the subjects to detect if any fractures re-emerge and if there is anything left of that traumatic residue that has characterised the first phase. From my empirical exemplification, I will demonstrate that *circular economy* seems to be still unconsciously spoken in the silos defining the EU as a regulatory, technical, and legal subjectivity and this poses a serious discursive limit to the picturing of the circle. Even when it is possible to depict a circle via the core issue of eco-design, this does not automatically describe a pattern of circularity that considers complexity, metabolism, regenerative capacity, or planet boundaries at large, as strong signifiers. Rather, it resembles a vicious circle of enjoyment whose pace and size must be able to cover a growing demand for goods. Hence, rather than being spoken of as a new centre around which our socioeconomic relationships must be re-thought, *circular economy* looks like another policy carriage of the EU as regulatory train, supported by the fantasies of standardising practices, increased recycling, eco-design, and new business models which promise to cover the impossibility of the discourse and a semblance of circularity. Nevertheless in the last part of the chapter, I shall focus on an important element of counter-resistance within the University-Capitalist social link, which can be attributed to proactive stakeholders such as the EESC, as well as to the unofficial policy landscape such as the Postgrowth Conference. These

DOI: 10.4324/9781003378228-7

seem to keep that original disruptive element of 'circularity' alive by expanding the circular economy's signification beyond simple recycling, energy efficiency, or the ever-expanding geographical boundaries, however their efficacy can strongly be hindered by the law of the dominant University and Capitalist social bonds.

Is it a circle or is it a line?

We will now look at how the idea of circularity is spoken by the EU in the 'acceptance' phase to investigate if there is any remainder of that element of disturbance that had characterised the first phase, the phase of resistance. The Lacanian discourse analysis conducted in the previous chapters demonstrated that a seemingly closed discourse is in fact open due to the presence of the inconsistent subject of the enunciation. As a result, we can observe the effects that *circular economy* has on the subjects observed and interviewed in terms of how any fractures in the discourse – perceivable in inconsistencies, gaps, and weaknesses – are produced, insofar as they demonstrate that *objet petit a* as an excess of sense remains, it is not neutralised. In the following empirical exemplifications, some DGs such as DG Energy associate *circular economy* with a series of standard labels to add to their product policy, some others such as DG Clima associate it with increased recycling an action at the periphery of the product life cycle, whereas those DGs involved more closely with the circular economy, namely the lead DGs DG Environment and DG Grow, seem to have an idea of the core of the circular economy, that is, acting on design. The next sections will therefore illustrate how the silos approach to work unintendedly returns in the speech, despite more integration and streamlining being advocated by DG Environment and DG Just.

Circularity as product labels

In the following extract, a speaker from DG Energy resorts to the metaphor of a train which is, I would suggest, emblematic as we can understand the signifying chain as the carriages of the train of language. In this case, we see the *circular economy* being compared to a carriage of a locomotive, where *circularity* is associated with standards, that is labels and letter scores, to be integrated and attached to a product policy:

> It's another example of the locomotive and the carriages. *So, the locomotive is the product policy,* and we have extremely *good product policy for energy efficiency and that's a label* on the fridges and the banning inefficient fridges and we have that for many products. *And it's easy for us to add other carriages to that train.* So, for a long time with washing machine and dishwashers *we've added water use for example, we are now studying how we might have a reparability label or a letter score,* the score on the label so your label in the shop wouldn't just tell you how much energy you use. But also, if it breaks down, how easy it is to get repaired. That's like another carriage on the train. We are, we've been looking into . . . how to

add durability requirements, recyclability *requirements as minimum requirements* for products.

(Interviewee 8, 14 November 2018)

It is worth noting that the metaphor of the locomotive and the carriages was mentioned a few times during the interview. Sometimes this metaphor stood for the Commission being the train and the DGs being its carriages, probably a metaphor for each DG representing a specific policy area. In this case, the locomotive stands for the product policy with the carriages being the standards, that is to say, the attached labels or letter scores referring to energy use, water use, and other elements of the circular economy such as reusability and reparability standards, which allow consumers to make informed choices when purchasing and consuming. Thus, in the DG Energy unit that is responsible for energy-related products, the image of *circular economy* is associated with standardisation via the labelling practice on reusability and repairability that comes with an energy product, with *circular economy* being another policy carriage to attach to the train. In the next section, the *circular economy* is associated with a policy of increased recycling.

Circular economy as recycling

Other examples indicate that *circular economy* is spoken of as more recycling, which can be pictured as an intervention at the periphery of the circle rather than at its core. Thus, we might wonder whether *circular economy* simply becomes a synonym for waste management and whether this helps break the discursive boundaries of the 'line' and depict instead 'circularity'. For example, DG Clima speaker explains below:

> So, *where do you get all the resources if you don't recycle more and better?* If we say we need to have electricity transports system *you need to have a way of recycling better which is not being done to the best extent today.* So, the circular economy is part of the solution very clearly and just look at the staff working documents next week, you will see a lot on circular economy.

(Interviewee 10, 21 November 2018)

This extract associates more explicitly *circular economy* with better recycling. The logic seems straightforward, reasonable, and even desirable: population grows, we cannot extract more and more new resources, but we can recycle more. Moreover, if we are not recycling properly, we can improve. On the other hand, we might wonder whether the strict focus on recycling is not enough for speaking of circularity in that it can still be based on the idea of fuelling a productive recycling industrial machine. During the UNFCCC COP24, at the EU side event 'Circular Economy, the missing link in climate action?' Yvon Slingenberg (DG Clima) claimed that 'the recycling industry has been mentioned has a major role to play' (11 December 2018). In this case, we might infer that, no matter how much we produce and consume, as long as we fuel an ever-growing recycling machine, we have circularity. For example, in the next extract from DG Clima, the speaker seems to be caught in the trap of the discourse:

> I think if you look at what I mean an element of what you say is always true. I think if you go back *even 10–15 years ago the idea of having a strong industrial complex to show more and more goods was a notion that more people even in Brussels would have defended. The thing that today we still want a very strong industrial complex we might need to do more recycling, more re-use, different business concepts that's all very different* . . . I mean the 2 related DGs on circular economy are DG growth and DG environment. So that is happening. They are really trying to talk to industry. Or talk about how *re-use and recycling can increase in Europe* or how we can be *resource efficient*. They are really doing that.
>
> (Interviewee 4, 11 October 2018)

The above extract began with the speaker stating that 10–15 years ago, having a strong industrial complex and ever-growing production and consumption would have been strongly defended even in the Brussels headquarters, leading us to guess that perhaps today things have changed. But shortly after, the speaker is caught in the chain of the discourse and a split between personal considerations and the constraints of the social link spoken through them is produced. This can be inferred when the speaker claims that 'we still want a very strong industrial complex' (Interviewee 4, 11 October 2018), which makes recycling necessary. Hence an intervention on the end-side of the product, that is recycling, is necessary to keep the premise unaltered, the Master Signifier (S1), with 'knowledge' on climate action and circular economy being mobilised accordingly. Hence, a *circular economy* as recycling can lead to the effect of having an ever-expanding recycling industry, which will be dictated by the same command of competitiveness and accumulation like any other industrial sector. Moreover, the meaning of *recycling* is contentious as well, in that behind this signifier lies a measurement rate that is at the centre of a dispute between Member States. Therefore, if for these speakers, *circular economy* means increased recycling, and more generally becomes a metaphor for 'recycling' or 'waste management', we might wonder whether this is enough to break the discursive boundaries of the 'line' and depict instead 'circularity'. At the same time other DGs, notably those involved with the circular economy more closely, seem to have a wider understanding of circularity and it is to the design component that we now turn.

Circular economy as eco-design

If DG Clima and DG Energy associated *circular economy* with its peripheral aspects or 'recycling', those DGs involved more closely with the circular economy, namely the lead DGs, DG Environment and DG Grow, seem to have an idea of the bigger picture, which involves acting on design. In their view, eco-design is a necessary component in the circular economy. In this respect, DG Grow explains:

Of course, what's happening now is a more active attempt in a range of poli-cies. How you can do that. For us, we see it in a battery what we're doing where *we want to make sure right from the start* that the requirement for bat-teries is that batteries in cars means that they can be easily recycled.

(Interviewee 11, 22 November 2018)

In this case, we see that in DG Grow the shift towards the regenerative component emerges spontaneously: 'Make sure right from the start that batteries are easily recyclable' (Interviewee 11, 22 November 2018). In fact, DG Grow explains the importance of eco-design, at least from the point of view of their own DG:

Eco-design is the core objective of this . . . colleagues would tell us that all of this can be recyclable and there are factories in Belgium that can recycle this. The key issue, and this is *what has to be tackled by design,* is a company that is doing the recycling does not want to receive a battery when a person spent an hour to undo all the screws and take out all the different elements. Because that's what makes the process unaffordable. *What you need is a bat-tery that can go into the machine and being dismantled automatically. So, you go to design if you can do that that shouldn't be mission impossible. Then recycling batteries is straightforward* and we can do with the technology we have at the moment and re-use all the core elements for recycling. So, of course the things with cars unlike with mobile phones, you will have people finishing with their cars then they go back to dealers they're not going to be thrown away in the bin, like mobile phones are. They should come back into the chain . . . as long as you solve this design issue not only you can have an obligation of manufacturing to ensure all the recycling in place and make it easy and profitable, but we'll see.

(Interviewee 11, 22 November 2018)

Thus, the problem of eco-design in car batteries is not the dismantling and reas-sembling tasks per se, as car batteries are already recyclable. The main issue is that this is a labour-intensive task, which as such is not perceived as a cost-efficient solution. On the other hand, the more desirable automatisation casts doubt over the increasing volume of car batteries to be recycled in the future. It also calls into question the development of technology and related technological advances that have their product life cycle. At the same time, bringing *eco-design* signification into the debate causes an interesting shift towards the signification space of energy efficiency that I have already addressed in depth in previous chapters of this book. In this regard, Yvon Slingenberg, Director of DG Clima at the COP24 side event 'Circular economy the missing link to climate action' (11 December 2018) claims:

This is also about product design. What we're trying to do there, at EU level, we have policies we call *eco-design we look at standard settings for products when it comes to energy use.*

(Yvon Slingenberg, 11 December 2018)

In fact, from this extract, it seems that *eco-design* is firstly signified along the energy efficiency signification space, perhaps as an effect of the co-optation of the *circular economy* into the hegemonic climate knowledge. For example, the Communication Eco-design working plan 2016–2019 (COM (2016) 773 final) claimed:

> *The Eco-design and energy labelling framework* has been one of the *most effective policy instruments at EU level to promote energy efficiency*, estimated to contribute around half of the energy savings target for 2020. The Eco-design and Energy Labelling legislative framework has the dual purpose of ensuring that *more energy-efficient products come to the market* (through eco-design) *while encouraging and empowering consumers to buy the most efficient products* based on useful information (through energy labelling).
>
> (COM (2016) 773 final, 2)

On a positive note, although the signification space in which *eco-design* was confined and stabilised at the time of the interview was that of energy efficiency, a section of the Eco-design working plan addressed the limits of eco-design qua energy efficiency and advocates a broader focus for the future:

> *The possibility to repair, remanufacture or recycle a product and its components and materials depends in large part on the initial design of the product*. It is therefore crucial that these aspects are taken into account when investigating possible Eco-design implementing measures. *The focus has so far been on improving the energy efficiency of products even if resource efficiency provisions* have been part of the Directive since its original adoption in 2005 and were introduced for some product groups with criteria concerning e.g., water use and durability. With this new working plan, the Commission will explore the possibility of establishing more product-specific and/or horizontal requirements *in areas such as durability* (e.g. minimum life-time of products or critical components), *reparability* (e.g. availability of spare parts and repair manuals, design for repair), *upgradeability, design for disassembly* (e.g. easy removal of certain components), *information* (e.g. marking of plastic parts) *and ease of reuse and recycling* (e.g. avoiding incompatible plastics), greenhouse gas and other emissions, and to further establish the scientific basis for developing corresponding criteria that meet the requirements of the Eco-design Directive.
>
> (COM (2016)773 final, 8–9)

As a result, although the eco-design signification was at the beginning confined to the signification space of energy efficiency, it seems to be slowly sliding into a new battery of signifiers of material flows, which indicates how the discourse is in fact never static or closed and that fractures can be leveraged upon. For example, the new Circular Economy Action Plan (2020) – another EU Commission Communication – embraced a wider notion of circularity by including Electronics and ICT, batteries and vehicles, packaging, plastics, textiles, construction and buildings, food, water, and nutrients (COM (2020) 98 final). At the same time, the next section suggests that the *circular economy* is being pushed and pulled along the signifying chain of allegedly new business models.

Circular economy as service-based model

New perspectives are constantly being added to the notion of *circular economy*. An interesting one is the service-based model, which did not necessarily originate from the philosophy of circularity but is today accepted as a part of the new idea of circularity. In fact, its narrative is perceived by some circular economy advocates, such as the EESC, as beyond the control:

> Because I've had a lot of problems with how the narrative is going . . . I'd also like to have a conversation on it . . . because it's *constantly changing and evolving. There are new perspectives on it. . . . One is that I think people don't distinguish and so it creates confusion you probably hear people talking about the service-based business model.* Are you familiar with the idea of selling a service rather than selling a product? Philips is always selling these examples. Caterpillar does a lot of work in a lot of different groups. And people would say what about food? How can you include food because we produce food, *we can't have a service model on food* we still need to produce food and either it's wasted, how do you do that?
>
> (Interviewee 5, 30 October 2018)

When speaking of *circular economy* as the service-based model we must consider that this new business model did not develop on a circular economy premise – at least in its original philosophy – but it originated elsewhere. 'Circularity' became attached to it at a later stage, like one of those carriages of the train mentioned by one of the DG Energy speakers. However, it can be argued that a change in ownership by selling light as a service rather than light bulbs does not automatically mean becoming circular. One reason for this is that not all goods and products in circulation can accommodate a service-based business model such as food, as the speaker explains. Second, assimilating an idea of circularity aiming to save resources into the service-based model creates the illusion of selling fewer products, in that, it involves a metaphorical shift from the tangible 'good' to the intangible 'service'. For example, if I buy a service rather than a product such as laundry services rather than owning a washing machine, it gives me the illusion that the company is selling fewer machines and they are making profits on the 'service'. However, if the planned obsolescence of the machine is not eliminated,[1] after a few years the company who owns the machine might force me to rent a brand-new machine, with the result that they are not selling fewer machines in the first place. The situation is comparable to those services we have on our mobile phones or laptops, such as apps or software, that after some time become incompatible with our devices, forcing us to replace them. Consequently, by embracing this signification shift of *circular economy* qua service-based economy we might create the illusion that the shift from the tangibility of 'goods' to the intangibility of 'services' will automatically benefit the climate and the environment. However, if we merely imagine substituting a tangible 'good' with 'service' by keeping the logic of increasing energy volumes and goods in circulation unaltered, the contribution of the service-based model should be further assessed. In this sense, the service-based model as an integral part of circularity would appear to be one of those Lacanian fantasies that the subjects desire to believe, whose suturing function is to promise to erase the impossibility of signification by providing the illusion of achieving a given desired object, where in truth these subjectivities are trapped in the same closed circuit of commodities, i.e. not tangible goods but services. More to the point, this illusionary sense of plenitude translates again into *jouissance* as a paradoxical and painful satisfaction, where production and consumption patterns remain virtually unaltered, sustained by the illusion of consuming services rather than material goods.

One circular economy or some circularity?

The above examples do not seem to suggest that *circular economy* can be regarded as the new core principle around which to reorganise our economy, namely that a linear economy will be converted and reconceived as a circular one. Adopting a circular economy within the EU, in fact, seems to imply only adding 'more circularity', as we shall see in the interview quoted below, which would be in line with the idea of adding another carriage policy to the train of the Commission. For example, when I asked DG Research and Innovation if the circular economy involves a reconversion of all economy, their answer was:

> *No, but more circularity we think is good* for climate, for the resources, for pollution, for many aspects, sustainability in the wider sense.
>
> (Interviewee 9, 16 November 2018)

The interview extract above suggests that the EU is not aiming at converting the whole economy into a circular one but at increasing the degree of circularity for a vague idea of sustainability. Similarly, during an interview with DG Energy, I tried to understand whether a perspective of circularity and material flow is borne in mind even if they work with such sectorial approaches. For example, I asked whether the bigger picture of the circular economy should always be considered. To prove my argument, I pointed at the wooden table at which we were both sitting, as the 'table' is not an energy-related product, but it is still part of a manufacturing process. The answer from DG Energy points again to the usual division of work and perhaps not to a transversal view (sic):

> No no, *it's not about everything because our train is called energy related products.* But when it's an energy related product, which means energy use product in practice, well for various technical reasons then we can do recyclability, then we can do reparability, we can do *other things. Our train is a*

legal train, it's a regulation train which has a particular content, it's never going to be DG Energy which proposes another train, another eco-design directive for tables. I wouldn't have any problem with that happening, *but it's just it would be bizarre for DG Energy to do it. But we happened to do it for energy use products and that has been very nice to bring circular economy aspects* as well. *And we prioritise energy, sorry, circular economy aspects which have direct benefits for consumers, and which can be measured on the product because our train is like that.* I mean, you know where we set energy requirement on a fridge, then there is a whole lot of (cannot be heard Ed.) things checks. By buying the fridge so you can check also whether the fridge has HFC in it, was HFC? Anyway, you know what I mean, or if it has met the reparability criteria by looking at the fridge. You can't check how much energy was used in manufacturing it, you can't see . . . what I mean, *our train, it's pretty good. But we can't do everything. But we like doing the things we can do. . . . We resisted these carriages to our train* but now we are convinced that is the right thing to do and we are happy to do it.

(Interviewee 8, 14 November 2018)

Circularity here refers though to those 'circular economy aspects' that can be added to the product policy, or energy-related products in that DG. A couple of interesting and interconnected aspects come with this extract. First, it still appears that another split between personal considerations and the constraints of the social link spoken through them is produced. In fact, the speaking subject qua EU representative as a neutral, quantifiable, and regulatory agent is still spoken by their silos and cannot depict a holistic approach encompassing all materiality (when I pointed at the wooden table): 'No no it's not about everything', 'It would be bizarre for DG energy to propose this' (Interviewee 8, 14 November 2018). Within this framework, the speaking subject mobilised the battery of signifiers available to them, but this act of identification with language failed to achieve the desired full representation, and once again the discourse seems to be fractured. It can be argued that the linguistic acts of identification within this regulatory role, defined as a 'regulation train' (Interviewee 8, 14 November 2018), and the deployment of skills and knowledge within this type of subjectivity prevented us from depicting circularity. Finally, towards the end of the extract the speaker betrayed a sense of dissatisfaction with the way in which the 'carriages' (of which the speaker is a representative) perform and in this we can appreciate an exemplification of Lacan's *jouissance* at work: 'Our train, it's pretty good. But we can't do everything. But we like doing the things we do' (Interviewee 8, 14 November 2018).

As it has been illustrated with the examples above, the most striking aspect is that the silos defining the EU as a regulatory and technical subjectivity return every time in the speech as these are the available discursive representations in their proximity. It can thus be argued that the way the circular economy knowledge is put to work can be a discursive limit to the picture of 'circularity', in that it translates again into some linear reductionist knowledge – which asserts itself as a neutral and objective – around circular economy and climate mitigation action more in general, in line with the co-optation of a new floating signifier into the hegemonic University discourse and Capitalist discourse. At the same time, even if we assume that all the aspects of circularity are considered, which might potentially lead us to see the 'circle' and no longer a 'line' – from the core of eco-design to the periphery of recycling – a potential breakdown in representation occurs. This breakdown of representation, which reveals the limits and impossibility of full signification, occurs every time issues of 'pace' and 'sizes' of the loop emerge in the speech. It is to these aspects that I now turn.

The size of the circle and the pace of its movement

Even when the *circular economy* is seen as a loop by those subjects who acknowledge the core issue of eco-design such as the DG Grow speaker previously discussed, this does not automatically translate into a signification of circularity that includes complexity, metabolism, regenerative capacity, let alone ecosystem cycles and planet boundaries, as strong signifiers. Even in those cases in which it is possible to depict a circle via the core issue of design, this circle resembles an endless loop running at an increased speed, and large enough to cover the growing demand for goods. As argued in the previous section, when functionaries discuss the 'pace' or the 'size' of the loops, a potential breakdown of representation occurs. Regarding the pace of the loop, during a 5-day stakeholder event organised by DG Grow called 'Raw Materials Week' (see Table A.2 in Appendix) a graph pointed to a recycling gap between critical raw materials recyclability and actual demand for that commodity, proving the point that we would still need primary raw materials extraction. That specific slide showed the hypothetical demand curve for a commodity X and a hypothetical product lifespan of 25 years and explained that if – for example – the demand for commodity X increases from 2 million tonnes per annum to 12 million tonnes, the recycling gap is 10 million tonnes because the supply of critical raw materials from secondary sources cannot meet the increased demand. The evidence of the recycling gap is also reported and thus confirmed in the long-term strategy staff working document (COM (2018) 773 final supporting analysis).[2] Consistently, a DG Grow speaker mentioned the recyclability of car batteries: the following extract reveals that the almost total recyclability of car batteries 'clashes' with an ever-growing demand for cars. Thus the car batteries 'circle' cannot be expected to cover the whole demand:

> Well, there is good news because with car batteries, *as long as they're made in a form that is easy to dismantle at the end you can recover nearly all 98 per cent-99 per cent of the original* raw materials, so you can more or less have a sort of virtuous circle of recycling. What you can have as a problem is that the selling of electric cars continues to grow dramatically. Now electric cars last longer than petrol diesel cars, so the batteries are not going to be recycled for another 10–15 years. *So, there is this gap where you're going to*

have an increased demand for batteries. But the raw materials are not coming back into the system you have to get them from elsewhere, from mines. Or if we can be much more efficient on recycling for example mobile phones apparently, at the moment, we only recycle something like 10 per cent of mobile phones. If we make 100 per cent there's a company here, a Belgian company, a big supplier of raw material that's the equivalent of 2 large mines for lithium and cobalt, *sorry I'm not sure there's particularly a conflict. If you're going to build cars, however green they are, there are going to be environmental issues full stop. So, as always, it's a plus or minus thing.* The policy of the Commission the policies of the Member States and China and the States is that electric cars are a better solution than petrol ones, far from being perfect. *There are also huge implications in terms of infrastructures to support the recharging* of cars.

(Interviewee 11, 22 November 2018)

Hence, we have a confirmation that the secondary raw materials market can cover only a small part of the demand. If the car industry grows, a shortage of recycled batteries to cover that car production in the short term can occur. This would mean engaging with new mining activity and further raw materials extraction. However, this does not seem to be particularly conflictual from the point-of-view of the representative of a DG whose aim is growth: 'Sorry, I'm not sure there's particularly a conflict' (Interviewee 11, 22 November 2018). In this case, it seems that an encounter with the traumatic point of signification, the 'return of the repressed' (Bettini, 2019; Hook, 2013; Lacan, 1988), that is the impossibility of having both a growing car industry and batteries circularity, is neutralised and positively integrated into signification to the point that it is not perceived as conflictual. Indeed, a sense of painful (dis)satisfaction, that is *jouissance*, seems to emerge within this impossibility: 'If you're going to build cars however green they are going to be environmental issues full stop. So as always, it's a plus or minus thing' (Interviewee 11, 22 November 2018). It appears that this state of full representation, embodied by a never-achieved state of 'circularity', is always deferred and the sense of plenitude of signification is only illusionary.

Thus, more circularity, does not necessarily mean an economy that goes at a slower pace, that flows slower than a linear one, or that respects the times necessary for a regeneration of the system. In fact, by keeping the demand and supply logic unaltered this translates into a fractured loop of *jouissance* that must go fast enough to cover the demand. Moreover, whenever circular solutions are not enough, the purpose of circularity is jeopardised because it simply means more extraction of raw materials. As a result, we might have small circles of recycled batteries within a still predominant linear economy that attempts to respond to increasing demand with increasing supply. The *circular economy*, therefore, becomes another fictitious change or transition that gives an illusionary sense of plenitude which in truth creates that paradoxical painful satisfaction that we call *jouissance*.

Similarly, a breakdown of representation seems to occur every time 'the size' of the loop emerges in speech. For example, the following speaker from DG Grow starts mentioning the eco-design and recyclability of car batteries, but then upholds the EU car industry competitiveness and world market expansions:

> So, one of the things by introducing standards is that *you encourage industry to invest more in innovation in this area so that they can remain competitive not only in the EU and meet the EU standards but also being competitive on the world market.* And there's no bigger market for cars than China so I mean I think I might get the years wrong but not that long ago the market in China, 10–15 years ago, was the same as France, something like 3 million cars a year. Now it's something like 23 million cars a year *so it's a huge, huge market which is important for the future of our industry.*
>
> (Interviewee 11, 22 November 2018)

This circle or circles seem to be embedded in a lead DG that looks on the one hand at eco-design issues and on the other hand at world market expansion where production is delocalised, and international shipping is the norm, affecting the way we picture circularity. At the same time, this signification twist should not be surprising if we consider that the *circular economy* is in the hands of DG Grow whose existence, subjectivity, is inherent to growth.

Eventually, I have illustrated that the fractures regarding *circular economy* re-emerge in the discourse even in the co-opted 'acceptance phase' and *circular economy* as a signifier is unable to confer the meaning that the entire chain offers. Instead it is signified in different forms and shapes, as a newly found climate policy, as product standards and related labels, as waste management, as eco-design qua energy efficiency (but potentially beyond), as big and fast loops of products' demand and supply, as a new business model, at different times of the enunciation. As with the case of *energy efficiency* and *renewables*, the main point here is not choosing what the 'real' meaning of *circular economy* is. Rather, starting from the signifier and how the split subject arises as a consequence of the signifier, the signifier *circular economy* is ultimately all that is being articulated – inconsistencies and gaps included – as these embody the excess to the discourse that sets desire in motion, namely *objet petit a*, and its corollary *jouissance*. The object cause of desire therefore is not completely neutralised, although its efficacy is drastically reduced. It is therefore concluded that the 'acceptance' phase has unsurprisingly entailed a loss of the traumatic potential that the 'circularity' metaphor brought with it, to the point that *circular economy* lost its very same peculiarity of 'circularity', thus the possibility of bringing a change of paradigm. In fact, the apparent circularity is translated into a new loop, a circle of endless (dis)satisfaction in which the subject is trapped. This new loop carries with it the painful pleasure of seemingly going more and more circular – through more recycling, more eco-design-energy efficiency, new business models, faster and bigger loops – without seeming to achieve the intended transformation desired for the sake of climate objectives. At the same time, we might wonder whether there is anything left of that original disruptive element that had caused resistance.

The element of counter-resistance

By making use of the Lacanian subject, I have so far shown where the fractures of the University-Capitalist circular economy social bond reside. I concluded that although *circular economy* potentially entails the expansion of the very narrow signification in which these DGs work regarding climate action (climate + energy nexus), simply annexing *circular economy* to a climate action qua energy metonymic slide does not automatically translate into a change of direction in terms of how climate mitigation action is thought and constituted. It can be observed that this expanded signification does not necessarily determine a picture of 'circularity' and we can still speak of a 'line' – that is a forward-moving regulatory train, policy after policy, silos by silos – rather than a common pattern across policy areas. As such, circularity can be compared to a fantasy in the Lacanian sense, that is, a socio-symbolic construction that promises to cover the subjects' lack. However, despite considering these fractures in the discourse and the neutralisation of their 'traumatic' potential into the dominant signification, we might wonder if there is any remainder of that disruptive element that caused resistance in the first place. This counter-resisting role, which in the theoretical and methods chapter has been called 'hystericizing', can be attributed to the stakeholders in the field such as the EESC who actively lobby the Commission, and to the researchers and activists that populate the unofficial policy landscape of the EU, such as those observed at the Postgrowth Conference. These subjects, by keeping a certain understanding of the *circular economy* in the agenda within a signification network in which it seems to have lost its traumatic potential, constitute forms of counter-resistance that attempt to re-introduce or keep the element of defiance alive. For this purpose, they push signification beyond recycling, beyond eco-design qua energy efficiency and resist the ever-expanding size of the loop by fighting against delocalisation and territorial dispersion of production and consumption.

The feeling of discursive struggle can be rendered by reporting an example from the unofficial policy landscape. At the Postgrowth Conference (18–19 September 2018), during a session called 'Technology, Sustainability and Growth, the Holy Trinity?' an ENGO representative asked the panel a quick question: 'Circular Economy, friend or foe?'. The question was addressed specifically to José Belver,[3] FUHEM Ecosocial, and the answer that the economist gave was the following:

> I would say, if circular economy is *another new greenwashing concept* just to make us, create an illusion that we can continue to have infinite growth *then it's clearly a foe of course.* But if we're talking about the *articulation between industries* that . . . the Commission has been talking about that, this is important sure. But not only that. *But if we're talking also about the adjustment to the ecosystem cycle and using mostly biodegradable elements* and materials then *it might be a friend. Also, together with sufficiency targets of course.*
>
> (José Belver, 18 September 2018)

This economist and researcher pointed to the fact that there is no such thing as 'circular economy' but that depends on the dominant signifier to which it is anchored and put to work. If it is anchored to the infinite growth command, that the speaker above defines as 'greenwashing', then it is perhaps not a friend. It can be argued that it would be a friend of the industry rather than of the environment. On the other hand, if it is built along an alternative new signifying chain of ecosystem cycle and planet boundaries, of sufficiency, it can contribute to climate objectives.

Similarly, an EESC stakeholder interviewed counter-resists the metaphorical shift of *circular economy* qua recycling and explains what the implications of focusing on recycling are:

> Recycling represents a loss of raw materials and a loss of energy. . . . For the first year the recycling industry would invite me to conferences . . . *and I would say if the circular economy is successful this spells the end of your industry. And they would be – No no this is our new future, circular economy, and recycling industry, perfect we're going to expand, get bigger.*
>
> (Interviewee 5, 30 October 2018)

Within this framework, one might wonder whether the newly advertised circular economy contribution to GHG emissions reductions needs to be further assessed. In fact, *circular economy* qua recycling does not mean that less of that material is produced in the first place. To better understand the difference, let's think, for example, of plastics in our everyday life. Would it be more beneficial to the environment to keep consuming plastics by assuming that they will be recycled, or would it be better to minimise their use in the first place? In the first scenario, plastics would enter a secondary raw material facility, that is the industrial process that recycles plastics. In the second scenario, some plastics are avoided altogether.

Furthermore, it seems that thinking of recycling as the actual amount of material that is re-used is misleading. In fact, when we think of 'recycling' we might instinctively think of what is re-used and what re-enters the productive cycle. In the extract below, the EESC interviewee explains that the recycling rate is calculated based on collection and not the actual re-use of materials as secondary raw material:

> So, at the moment you have material being used in the economy . . . they're collected by waste management companies, and they go to the waste management facility and *when they arrive at the waste management facility there's a certain volume . . . and then they go through a sorting process within the facility.* . . . When you come out of the end of the waste management facility the volume that comes in was this much (indicates, ed.) and then when you come out is this size (indicate a smaller size, ed.) you know because you've lost some, understandable, this the process. *This smaller amount is the valuable material that's available as secondary raw material for other industries to use. At the moment, recycling rates are measured across Europe mostly back here before they get sorted out.* Ok, so back at this stage they go . . . your

recycling is this amount 10. But the reality is 10 in (indicates their country of origin Ed.) for example 10 tons arrive at facility but when we look at what is sorted out it's maybe 5 tons, you lose half. So, *the Member States understandably were fighting for their own interests saying currently according to the statistics our recycling rates are at for example 50 per cent and we agreed to these targets of 65 per cent because we can achieve that.* If we measured that here which is not the true measurement really what is being recycled. It's the measurement of what is being collected and so the argument was do we measure here, or do we measure here? if we measure further along . . . ok so they call it pre-sorting early stage and obviously afterwards post sorting measurement. Obviously suddenly all the Member States would have to remeasure, and their starting point would drop. They would go we're doing way worse than we thought we were doing *and then in truth we're doing way worse than we were pretending we were doing. Because pre-sorting figures are actually irrelevant, you know, they tell us something about the collection which is important a good collection. They don't tell you what you're contributing to post sorting in the second raw materials markets.*

(Interviewee 5, 30 October 2018)

From this extract, we understand that a potential breakdown in representation occurs every time we associate *circular economy* with waste management. This can be explained by the way in which our knowledge of recycling rates is produced and by the meaning that lies behind that measurement, that is, pre-sorting waste collection, as opposed to post-sorting recycling as secondary raw materials.[4]

More importantly, the *circular economy* as recycling, regardless of how it is measured, does not automatically close the loop, and depict circularity, as signification still gravitates around the endpoint of a product life cycle. In fact, although one might argue that to have recycling something must have been made recyclable in the first place and that interventions on the design side are implicitly given, the EESC speaker explains how recycling in the concept of circular economy is peripheral. More importantly, very high recycling rates are still compatible with a linear economy (sic):

Back at the design, which is so important to create the transformation, the motivation for the waste management facility to be restructured, to become secondary raw material preparation facility. Not waste management. So, our own philosophy on circular economy is it's not about waste. At the moment the EU is *achieving high standards on circular economy*? Not really, it's achieving high standards on waste management . . . recycling is peripheral, you know it's not at the core of circular economy . . . *more recycling is phase one.* But then quickly you need to move to *phase 2 which is beyond recycling,* you know *driven by eco-design,* to remanufacturing, to repairing, to reusing, not recycling. Especially for *short shelf-life goods recycling very quickly represents a degrading of the virgin raw material.* You know, for aluminium Coca Cola cans in 2 weeks they go from being extracted to be on a shelf and even if you have 95 per cent recycling rate, you're losing 5 per cent of the raw material every two weeks, very quickly your virgin raw material is depleting *and it's still a linear model. So, this is an example of even 95 per cent recycling rates still promoting the idea of a linear model.*

(Interviewee 5, 30 October 2018)[5]

If the *circular economy* as recycling is only a tiny peripheral part of *circular economy*, and if this is still compatible with the linear model, the core of circularity lies in its focus on the departure point of the circle, the eco-design component. This is what defines the basis of a circular economy, perhaps because this is what makes the regenerative component possible. In the following extract, the EESC speaker spontaneously – with no intervention from the interviewer – resists the closure of this signification space by saying how energy efficiency is only one tiny part of eco-design:

The Circular Economy Action Plan,[6] the vision identified a number of actions . . . and the Commission has this chart of what happened and what has not happened . . . like a spreadsheet and the ones that are achieved are green and 80 per cent are green so it looks like they're delivering on everything. But the details is that, for example, one of the actions is eco-design, developing Eco-design working plan, they did it and again because this is a new communication and the working plan, it has to go through a process and again here . . . but the *Eco-design working plan only focus mostly focus on energy efficiency of products whereas for circular economy eco-design needs to include tables and you know it needs to include everything ehm buildings, products* . . . whereas the Commission's eco-design focus is pretty much on making sure the product is energy efficient which is *one tiny part of eco-design.*

(Interviewee 5, 30 October 2018)

However, I have shown that while initially defined in the signification space of energy efficiency, the new Circular Economy Action Plan made important steps in this regard. Finally, the EESC stakeholder below seems to counter-resist once again the globalisation push of *circular economy* by addressing the importance of its local dimension:

So instead of a product being manufactured from component parts, from raw materials that are collected from all over the world, I mean that's our current model, I mean you collect all the raw materials, you don't pay people who are working in the mines properly to extract them, often slave labour comes into this, you get the materials from conflict areas. The phones are very great examples of it, you get the materials and then it's assembled in a factory somewhere and it goes back around the world to be packaged, and maybe back around to be rebranded as Sony, and then back around to be shipped to these different warehouses and then distributed. You know sometimes products go around the world 2 or 3 times before they come to me as a consumer at the end of it. *Circular economy eliminates that, ok? It takes the products that are currently in circulation and then it says we will promote business initiatives that can take these back.* And instead of being a waste management facility, it is a refurbishment facility, or it's a remanufacturing facility or it's taking apart products to get back to their valuable raw materials, which can only be done in an economically viable way if the eco-design is implemented before that. And once it's implemented it gives encouragement to the bigger industries who have deposits around the world and different localities to say – Ok we will invest in eco-design for our product. But then we don't want to lose the valuable materials so that it incentivises to have this new ownership model. When I finish with my phone Sony they will say – We take that back, please. To give you convenience we're going to a Sony take back centre in . . . where you live, and I can go – Here you go that's cheaper for me. Or get money back for it. And so, it will incentivize to have a local facility and people can work there disassembling, creating valuable secondary raw materials that actually have a value. *So, in that sense that kind of implementation gives power and autonomy to smaller centres.*

(Interviewee 5, 30 October 2018)

This extract illustrated that the subjects above resisted the ever-expanding size of the loop by fighting against delocalisation and territorial dispersion of production and consumption. However, the efficacy of these hystericizing acts, whereby subjects 'disturb' signification by challenging knowledge assumptions, can be ultimately hindered by the limits of the University and Capitalist social bond. This is in line again with the argument that the 'hysteric' can function within a University discourse, as an activist, a stakeholder, or as a civil society representative in this case, but their efficacy is limited and operates within the boundaries allowed by that specific discourse (Fink, 1999, 30).

In conclusion, this exercise has not been a critique of the circular economy per se: rather the critique is addressed as to how 'circularity' is spoken by its actors in relation to delivering an ambitious climate mitigation action. To summarise the story of the circular economy in the EU's climate action, we can conclude that after an initial phase of 'resistance' and in which the link to climate action was not as prominent, *circular economy* has been accepted into the EU's climate action through push and pulls across existing signifying chains and acquired relative stability in signification. *Circular economy* could have potentially entailed an expansion of the narrow signification of the EU's climate policy, beyond the climate + energy nexus and towards more holistic considerations of the chain of production and consumption interacting with the surrounding biophysical environment. However, it is still unconsciously spoken in the silos that define the regulatory technical subjectivity of the EU, notably in relation to a big and fast vicious circle. This means that rather than being spoken of as a new centre, a new anchor that would leverage a shift around which our socioeconomic and thus climate relationships should be re-thought or re-organised, or a common pattern across policy areas, the *circular economy* looks like another policy carriage in the hands of a couple of lead DGs that define the subjectivity of the EU qua regulatory train. This illusion of circularity is supported by the fantasies of standardising practice, recyclable products such as car batteries, eco-design qua energy efficiency, new business models, faster loops and bigger loops that promise to cover the impossibility of the discourse and provide instead a semblance of circularity. Hence, this phase of 'acceptance' has entailed a loss of the traumatic potential that the 'circularity' metaphor brought with it, to the extent that the *circular economy* lost its very same peculiarity of 'circularity'. More specifically, the apparent circularity is translated into a new loop, a circle of endless (dis)satisfaction in which the subject is trapped, which carries with it the painful pleasure – that is, *jouissance* – of seemingly going more circular without being able to subvert the hegemonic discourse and pave the way for the real transformation to pursue the desired and required climate objectives.

At the same time, elements of counter-resistance are present in the field and attempt to re-introduce or keep the original disruptive force of the *circular economy* alive and push signification beyond recycling, beyond eco-design qua energy efficiency, beyond the ever-expanding geographical boundaries of trade, and these might ideally contribute to circular economy being a real discourse of change. However, these 'hysterics' can function within a dominant university-social link, but their efficacy is affected by the law of the dominant social bond. Ultimately, it is impossible to predict whether a different mode of working will characterise any new Commission configuration in terms of circular economy implementation, or whether more radical forms of resistance by stakeholders and activists will eventually manage to open an irreversible fracture in the discourse.

Conclusion

In this chapter, I continued the story and followed the reintroduction of the circular economy within the hegemonic climate action discourse, which I called the 'acceptance' phase. I illustrated how the reintroduction of *circular economy* entailed a loss of the traumatic but liberating potential carried by 'circularity', which might have produced a real change, and how the EU's *circular economy* seems to be unsurprisingly co-opted in the University and Capitalist social bond of knowledge and *jouissance*, respectively. Nevertheless, through the Lacanian subject of the enunciation, I have illustrated that the discourse is fractured, and I have observed the effects of the *circular economy* on the subjects willing to bring effective climate action to detect whether discursive ruptures still emerge. Despite an ideal state of 'circularity', rather than being signified as a new core principle around which our socioeconomic relationships have to be re-thought, or a common pattern across policy areas, the *circular economy* looks like another policy carriage of the EU qua regulatory train and is still unconsciously spoken in the silos defining the EU as a regulatory technical subjectivity. This policy carriage is supported by the fantasies of standardising practice, recycling role, eco-design as energy efficiency, and new business models that promise to cover the impossibility of the discourse and give us a semblance of becoming circular. Hence, this 'siloed' picture of the circular economy poses a serious discursive limit to the picturing of the circle and even when it is possible to depict a circle via the core issue of eco-design, the circle resonates with an endless loop that must run fast enough and big enough to cover the growing demand of goods and services. Therefore, the depicted circle resembles more a loop of *jouissance* which reveals the impossibility of the discourse, and powerfully re-emerges whenever the speakers are caught in their discourse and betray a sense of doubt over pursuing their desired circular transformation. An important element of counter-resistance to this co-optation can be found in the social bond and is attributed to stakeholders such as the EESC or in the unofficial policy landscape such as the Postgrowth Conference, which seem to keep the element of defiance visible. These subjects indeed expand *circular economy* signification beyond recycling, beyond energy efficiency, and beyond the ever-expanding geographical boundaries. It should also be considered that the efficacy of these 'hysterical interventions' can be hindered by the laws of the dominant social bond, but the progress made in the 2020 Circular Economy Action Plan with regard to eco-design beyond energy efficiency represents an important advancement in this respect. Ultimately, it is impossible to predict whether a different mode of working will characterise a new Commission and Parliament or whether other stronger forms of resistance will eventually manage to open an irreversible, painful and liberating fracture in the discourse and how the *circular economy* will eventually be implemented. The next chapter will draw the main conclusions of this work and discuss its broader implications.

Notes

1 See the description of the event 'Tackling premature obsolescence in Europe' in Table A.2 in the Appendix.
2 'The EU is at the forefront of the circular economy and increasing the use of secondary raw materials. For example, recycling rates of some metals such as iron, aluminium, zinc, chromium, or platinum already reach over 50%. For others, especially those needed in renewable energy or high-tech applications such as rare earths, gallium, and indium secondary production represents only a marginal contribution. Significant amounts of resources leave Europe in the form of waste and scrap, which are potentially recyclable into secondary raw materials. However, given the scale of fast-growing material demand, primary raw materials will continue to provide a large part of the demand. Also, due to the long time spans until these reach their end-of-life stage, recycling opportunities will fully materialize with a lag of several years or, in the case of buildings, several decades' (COM (2018) 773 final support analysis, 259).
3 Researcher at FUHEM Ecosocial, Member of the Transitions Forum and the Inclusive Economy Group Economist. The EU Commission representatives Paul Hodson, DG Energy, Energy Efficiency Unit and Doris Schroecker, DG Industrial Technologies, Research and Innovation, Head of Strategy Unit were also present at the panel.
4 In fact, the second Communication on the circular economy cites:

'To raise levels of high-quality recycling, improvements are needed in waste collection and sorting [. . .] The revised waste proposals will also address key issues relating to the calculation of recycling rates. This is essential to ensure comparable, high-quality statistics across the EU, and to simplify the current system and encourage higher rates of effective recycling for separately collected waste'

(COM (2015) 614 final, 9)

At the end, the interviewee says that the process has moved further along, but it is not clear if, from the Member States, if this is going to be a post sorting measurement.

5 See also 'Circular economy research and innovation- Connecting economic and environmental gains' section 'Closed-loop manufacturing systems, 16'. This booklet aims to showcase how several areas of R&I policy are supporting the transition to a circular economy. But the disclaimer in the front-page states that this does not necessarily reflect the views of the EU Commission, but those of the authors (not listed).
6 The speaker here referred to the first Circular Economy Action Plan (2015) not the latest 2020 Circular Economy Action Plan.

8 Conclusions

The EU's climate action: a transformative discourse?

In summary, how can subjects disrupt and possibly subvert the status quo? What, conversely, prevents change and how can we distinguish real transformation from an apparent and fictitious change? Ultimately, is it possible to bring about a system change?

To answer these questions, I have developed a framework for understanding the EU's climate action and policy-making in a period that coincides with the finalisation of their 2030 medium-term strategy, and the launch of their 2050 long-term decarbonisation strategy: this corresponds to the phase of policy-making that immediately precedes the famous EU Green Deal, and therefore constitutes a snapshot of a wider and longer policy-making process. The period under observation has been referred throughout this work as the moment 'in between strategies' to emphasise the transitional aspect that this policy-making should bring about in the achievement of its climate mitigation objectives, in line with the Paris Agreement commitments. Building on Lacan's theory of discourse and via a Lacanian discourse analysis conducted on mitigation knowledge such as *efficiency*, *renewables*, and *circular economy* I have assessed whether these are able to bring about a system change and to what extent the speaking subjects observed are able to disrupt and possibly subvert the status quo. Further, the theoretical framework provided by Lacan – which I turned into an analytical framework – provided a methodological tool to investigate change, that is to distinguish real transformation from fictitious change and to identify what, conversely, prevents change. This concluding chapter summarises this book's main arguments as well as discusses its broader relevance in the field of GEP.

By applying Lacan's theory of discourse, I have illustrated how the EU's climate action as a multi-institution, multi-stakeholder and multi-level process is seen as the set of societal relations made possible by the discourse as an overarching linguistic structure, which constitutes the subjects' presupposed knowledge – 'what needs to be done' – to formulate and implement climate policies as policy-makers, stakeholders, citizens. At the same time the EU representatives, stakeholders as well as citizens as Lacanian political subjects are the (split) speaking beings who in the pursuit of given climate objectives are constituted through the available battery

DOI: 10.4324/9781003378228-8

of signifiers determining the climate relations available to them. As a result, the subjectivities produced are differently 'split' between those who are convinced by the discourse, those who acknowledge its breaking points but attempt to disavow these so that they are not perceived as conflictual, and those that overtly challenge the status quo knowledge, what we called the 'hysterics' in the field.

Via a Lacanian discourse analysis, I have demonstrated that it is possible to open a seemingly closed discourse and reflect on any possibility of disrupting the hegemonic discourse and deliver change. In this regard, the key role played by *energy efficiency*, *renewables*, and *circular economy* in the transition is emblematic because they do not resonate with the 'jam tomorrow' problem – typical instead of big engineering solutions such as GHGs removal technologies – in which the fantasy of techno-rationality sustains the promise of full realisation of climate objectives. Rather *energy efficiency*, *renewables*, and *circular economy* are instead desirable and implementable solutions in that no one would argue against having equipment or devices that consume less energy, use renewable energy or are re-manufactured.

However, we cannot think of *energy efficiency* or *renewables* or *circular economy* as a presupposition of sense that automatically delivers the desired emissions reductions, in line with climate science. Rather, by starting from an analysis of the unfolding signifying chain and its effect on the speaking subjects, I empirically detected how *energy efficiency*, *renewables* and *circular economy* are spoken by its speaking subjects and what becoming efficient, renewables and circular means for the EU in relation to delivering an ambitious climate change mitigation. The main conclusion is that the EU's climate mitigation action under observation appears less a real transition than an apparent and fictitious change and can be understood with the following attributes, where each aspect builds on the previous one.

The disavowed authority of knowledge

One of the main lines of investigation of this book concerned the type of knowledge underpinning the EU's climate mitigation action. In this respect, I illustrated how climate change knowledge is heralded by policy-makers as the ultimate authority and as evidence-based. Accordingly, a whole knowledge apparatus is mobilised and expressed through apparently rational and objective signifiers such as 'quantitative data', 'analytical recommendations' which set the overall direction to climate mitigation action. This pattern emerged when *energy efficiency* or *renewables* as climate mitigation 'knowledge' at work were immediately spoken and thus thought of in terms of targets and measurements and higher ambition is immediately associated with an increase in targets. This way the EU – as a technical and regulatory subjectivity – emerged as a facilitator of quantifiable and rationalised knowledge which is mutually reinforced by a horizontal and vertical bureaucratic organisation of work across policy sectors.

At the same time, these quantifications expressed a more complex relationship between climate change and climate knowledge production. Although the acknowledgement of the non-neutrality role of science in policy-making is in line with previous early studies on discourse in environmental politics, such as Litfin's research into ozone discourses (1994), a Lacanian critique made it possible to zoom in on the allegedly neutral attributes of 'measured', 'quantified', and 'rational' knowledge. As the EU put climate knowledge to work with the signifiers available to them, this overarching structure exerts power in a way that any alternative form of thinking that deviates from measurement, rationalisation, bureaucratisation, and valorisation becomes impossible and is discarded, with the most emblematic case illustrated being that of the circular economy. *Circular economy* was at first resisted from within the EU Commission by DG Clima and DG Energy which usually deal with climate policy in modelling terms, due to the difficulty in translating it into that quantified element which governs our societal relations. As a result, its reintroduction after the initial resistance phase has entailed a loss of its traumatic potential and its valorisation into the dominant University discourse of disavowed knowledge and Capitalist discourse in which knowledge becomes a means of *jouissance*. In other words, although *circular economy* could have potentially become a common theme across policy sectors, it has eventually been co-opted into the hegemonic discourse and is spoken in the silos defining the regulatory and technical subjectivity of the EU, which ultimately poses a discursive limit to the picturing a 'circle'.

Hence, the debate here is not on the deliberate politicised role of climate science, but on what we called the 'disavowal' of the performative and political dimension of a type of knowledge that looks and asserts itself as factual (Žižek, 2004, 394). In this regard, I emphasised how this element of disavowal is constantly at play: for example, by introducing a defying element such as 'imposing' any collective change of behaviour for the sake of climate goals, a sort of intolerance for a 'command' emerged and authoritarianism was evoked. In fact, it appeared that the resistance towards any tougher approach for the sake of climate goals in the name of freedom and democracy hides in fact a disavowed command, which emerged during interviews and events in the alleged factual role accorded to the 'consumer'. This imperative constitutes the real anchoring point of representation, its Master Signifier (S1), and therefore the logic limit of signification and suggests that we are subjected to orders. However, as it is often spoken of as factual and referred to as 'real world constraints' (Interviewee 8, 14 November 2018) or 'real life' (Interviewee 4, 12 October 2018), this form of order should remain unquestioned. As a result, climate change *mitigation* embodies Lacan's University discourse as it is spoken of and welcomed as a discourse of neutral, quantifiable and objective knowledge (S2) which is in fact commanded by an authority, the Master Signifier (S1), that sets the direction of the discourse qua social bond and regulates our intersubjective societal and climate relations. In the case of the objects of knowledge analysed in this book, efficiency, renewability, and circularity, these are still made meaningful within the old logical boundary of the discourse, evidenced by the pervasive presence of 'growing competitiveness' and its corollary signification chain of accumulation and ultimately economic growth which still fix the semantic ambiguity of the discourse. However, this disavowed Master Signifier is what makes the social bond possible, that element which decrees our existence as social beings, what confers apparent meaning stability to the signifying chain and to all related discursive practices governing policy-making and what ultimately provides the illusion of the real life and its governability. As a result, *climate change* and *mitigation* are spoken of and acquire meaning stability within the boundary of alleged evidence-based knowledge which is in fact commanded by the Master Signifier of competitiveness and accumulation. The implication is that 'climate change' and 'mitigation' resulted in the rationalised and reductionist knowledge of climate action as energy transition; *energy efficiency, renewables,* and more recently *circular economy* as a policy carriage, constitute therefore a by-product of this apparently neutral and factual knowledge which should be implemented not per se but in function of the modernisation and competitiveness of the EU economy and industry.

More specifically, my LDA demonstrated that the Master Signifier operates by enabling the juxtaposition of climate action with energy transition via the emphasis on the untouchable role of the subject-consumer in a low-carbon future and the so-called clean energy transition. Hence, the constitution of the EU's climate action-energy transition appears to be a cosmetic re-styling of previous sustainable development narratives, which have been studied under the lenses of ecological modernisation (see Hajer, 1995; Hulme, 2008). Accordingly, these objects of identification that should deliver the transition are the quintessence of the knowledge qua quantifiable rationalised, bureaucratised, reductionist entity at work. Notably, in the case of circular economy I emphasised how this is still unconsciously spoken in the reductionist and bureaucratised terms that define the technical and regulatory subjectivity of the EU, to the extent that this poses a discursive limit to the very same depiction of 'circularity' which gets partially lost in the enunciation. In this respect circular economy reductively seems only another policy carriage of the EU qua regulatory train rather than a common pattern across policy areas or a new centre around which our socioeconomic relationships should be re-configured.

The fractures in the discourse as an effect of the subject of the enunciation

At the same time, despite the apparent stability of the social bond, we should not think of the discourse as a closed structure: rather, it is in fact open and fractured and this can be explained by the intervention of the Lacanian subject of the enunciation who is produced by language as surplus of sense. Opening -up a seemingly closed discourse through a Lacanian discourse analysis made it possible to zoom in on the mobilised climate power-knowledge relationship and identify its points or rupture – the leftover between what the subject wants to say and what the subject says – which is rendered visible in the inconsistencies, gaps and blind spots that characterise the subject's enunciation. This is the only way in which we can expose *objet petit a* as the Real of the discourse, insofar as it escapes language and easy representation. However, this can approximately be compared to an ideal state of wholeness characterised by a harmonious, prosperous, safe planet, by infinite energy supply, by efficient and infinite clean resources and by perfect circularity, that which was lost (yet never-achieved) and needs to be recaptured. This is key, because *objet petit a* as (libidinal) remainder of signification sets desire in motion and thus carries a liberating potential that can either disrupt signification and generate the conditions for an alternative signification or, on the other hand, be positively integrated into signification. In practice, I have illustrated that the subjects observed and interviewed are split between the linguistic structures defining them and organising their climate relationships and the desire to come up with effective climate action and carry the ultimate transition, as they speak of a 'system change', of a 'new society' (Interviewee 9, 16 November 2018).

Consequently, I have exposed the nonsensical character of the signifiers that are the carriers of this transition to assess how this remainder is dealt with by the speaking subjects, in which 'nonsensical' does not mean that there are no GHG emissions to curb, no renewable energy, or no circularity to implement, but that they reveal each time their partial, never-achieved character when spoken in the enunciation. For instance, I have illustrated that *climate change* as a biophysical phenomenon being signified in terms of energy policy, as reflected in the division of work of the lead DG Clima and DG Energy, entailed a 'closure' of the signification space that excluded wider considerations about biodiversity loss, ecosystem damage, pollution, and the planet in its wider understanding. This residue of signification within the climate + energy nexus is however called into question by the wider and perhaps more progressive stance of DG Environment who tried to keep signification open to all the other environmental aspects and warned about those political solutions that can have other disastrous environmental consequences. As a result, *climate change mitigation* took shape as a socio-political reality in the narrower signification space of climate + energy and the same partial and inconsistent character can be highlighted in the case of *mitigation,* whose meaning was perhaps dependent on its institutional character and the silos' signification spaces. These signification spaces are in turn an expression of the bureaucratisation of knowledge and organisation of work. For example, the discussion around what defines *mitigation,* that emerged spontaneously within DG Energy, created some contradictions as to whether the meaning of mitigation should include renewables as well. I perceived that the distinction between *mitigation* and *renewables* originated in the speaker's association with 'targets', which links back to the pattern of quantification and rationalisation that characterises the climate action social bond. In the empirical section, I also illustrated that the speaker perceived *emissions reductions* and *renewables* as two different entities because they correspond to two different targets and they associated *mitigation* with the former targets (emissions reductions) but 'not so much' (Interviewee 6, 7 November 2018) with the latter targets (renewables). Moreover, as the interview continued, the meaning of *mitigation* became dependent on its institutional character, 'all the activities linked to DG Clima' (Interviewee 6, 7 November 2018) which demonstrated how the 'silos' approach to climate knowledge repeatedly returned in the speech although a different speaker asserted they work 'less and less in separated silos' and that 'things have changed a lot. . . . Not for the simpler as it has made everything more complex, but for the better' (Interviewee 9, 16 November 2018). The initial 'target' distinction was rendered more inconsistent as the speaker contradicted themselves by saying that GHG reductions (thus their mitigation) from that specific DG Energy unit's point of view is Renewables and CCS. Thus, whereas at the beginning of the conversation *renewables* was 'not so much mitigation', by the end of the conversation *renewables* became 'a sub-part of mitigation' (Interviewee 6, 7 November 2018).

The extracts from participant observations and interviews have illustrated that each signifier individually, whether *energy efficiency, renewables,* or *circular economy* cannot confer the meaning that the entire chain offers but are signified at different times of enunciations in different forms and shapes and full meaning is always deferred. For example, *energy efficiency* and *renewables* are spoken of as metaphors for legislative acts, as equations, as technologies and as economic savings in the case of *energy efficiency,* or as a technology rather than an energy source (or sometimes both) in the case of *renewables. Circular economy* as well is spoken in different forms and shapes at different times of the enunciation: the examples provided in this book pointed to a newly established link between *circular economy* and climate mitigation policy, whereas *circular economy* was previously associated and signified within the semantic field of resource scarcity. Within this new association to climate knowledge, circular economy was articulated as standards, as product labels, as better waste management, as eco-design qua energy efficiency, as big and fast loops dealing with an increased demand of goods, and as a new business model substituting the tangible character of the good with the intangible character of the 'service'.

The argument I have made is that opening and juxtaposing all the possible understandings, which are not necessarily consistent with each other, does not reveal any hidden meaning and establish an absolute truth as to what *energy efficiency, renewables,* or *circular economy* are. Rather, *energy efficiency, renewables,* and *circular economy* are all that is being spoken of, including the inconsistencies and gaps that arise, because it is these inconsistencies that embody the ontological surplus-loss of signification, insofar as *objet petit a* can only be accessed by disrupted signification.

Dealing with objet petit a, jouissance, *and the role of fantasies*

Despite the impossibility of the discourse, by resorting to *energy efficiency, renewables,* and *circular economy* as a presupposition of sense, as what needs to be done to deliver emissions reductions, we establish, understand, and maintain our socio-political climate relationships between policy-makers, stakeholders, and citizens. In fact, discourse and subject mutually presuppose each other, so that the real subject presupposes symbolisation (language) to create sense and, more widely, to establish and maintain these societal climate relations, but the discourse presupposes a real speaking subject to be established and maintained. Accordingly, these subjectivities set their climate targets, formulate their impact assessments and cost-benefit analysis, run official and unofficial consultations, engage with modelling activities, configure scenarios, release official policy documents, fund technology innovation and research and impart and enact 'green' or 'smart' consumption. Indeed, the subjectivities involved in the climate social bond are not solely the EU bureaucrats caught in their chain of consensus-seeking and technical-regulatory practices. They can also be the stakeholders negotiating their targets with the policy-makers, but also the new green jobs' graduates, as well as all the new citizens-consumers placed at the heart of this transition engaging with their smart consumption practices in a flexible decentralised smart energy market.

Overall, I have shown that this excess, residue of signification can either disrupt signification and produce alternative significations, and thus change, or can be positively integrated into signification as commodified knowledge or as consumption objects. Yet, ultimately, from the desiring position of having an effective climate action and in the pursuit of achieving climate mitigation as a form of ideal and ultimate enjoyment, the EU mobilised its presupposed knowledge apparatus by resorting to the battery of signifiers available to them and this in turn activates and fuels a double mechanism of fantasy at the socio-symbolic level and *jouissance* at the level of the Real. In this regard, I provided examples of how the traumatic points of the discourse are integrated into the status quo signification as commodified knowledge or objects, such as those cases in which a split between personal considerations and the EU representative from which they speak emerged. In these cases, the subject seems to be aware that their discourse is characterised by fractures and points in which representation breaks down, but there is an attempt to disavow personal considerations and turn these discursive holes in a positive feature, to the point that these are no longer perceived as conflictual. For example, the so called 'rebound effect' as a side effect of energy efficiency measures was initially denied by the IEA and then turned into something positive and commodifiable good for growth. I have provided a further example in which a DG Grow representative did not perceive having both a growing electric car industry and a secondary circular market of batteries as conflictual, in that according to the speaker's logic we can have both a growing electric car industry and some circularity of batteries. Hence in these cases, we can appreciate the relation between fantasy and *jouissance*: representatives are supported by fantasies as socio-symbolic constructions that promise to cover the impossibility of the discourse and achieve a desired object such as 'more circularity'. At the same time these representatives are caught in the chain of signification, in a closed circuit of commodities that provide instead a semblance of efficiency and circularity where we have commodified knowledge formalised by pieces of modelling, or consumption objects such as a set amount of electric cars.

Therefore, the impossibility of the discourse resulted in a constant and illusionary sense of plenitude, a paradoxical sense of (dis)satisfaction – that is *jouissance* – which emerged every time the speaker attempted to re-capture their (allegedly) lost object, namely every time the signifying chain of efficient buildings, efficient heating and cooling systems, efficient cars, more sophisticated modelling, more renewable technologies, better standardisation, better waste management, new business models, better bureaucratic organisation and coordination, is deployed and conveyed the sense that full satisfaction was always delayed. As it has been illustrated in depth in the analysis chapters, they do not seem to attain the ultimate full enjoyment of a real efficient, renewable, or circular change and thus the achievement of the desired climate objectives is endlessly deferred.

The summary provided so far justifies the initial argument based on which the acclaimed climate action qua energy transition appears to be an illusionary or fictitious change, a redistribution of tasks within the same old social bond, under the authority of a more 'efficient', 'renewable', and 'circular' Master. For this reason, I concluded that the EU's climate mitigation action under observation appears to be the formalisation of Lacan's University discourse and its complementary Capitalist discourse of enjoyment. The reason for that lies in the emphasis on (commanded) knowledge and for the fallacious sense of (dis)satisfaction and plenitude (*jouissance*) associated with it, sustained by the integration into the dominant signification of its traumatic points. These traumatic points however do not disappear, and this is why the discourse is ever partial and inconsistent. However, their transformative potential is drastically reduced and is unable to bring a revolution of the discourse, that is a shift to the Hysteric's discourse or the Analyst's discourse, as these breaking points of the discourse are positively integrated into the hegemonic signification and turned into commodified knowledge or objects. As a result, the possibility of experiencing their disruptive force is limited, and this ultimately hinders the possibility of transformation.

The dimension of counter-resistance and defiance

Although changes in language and in the discourse do not happen overnight, the Lacanian approach adopted in this book enabled us to understand where fractures in the discourse are located and how to leverage on these to produce an alternative discourse. For example, I demonstrated that although a simple introduction of a new signifier such as *circular economy* into a dominant signifying chain does not result in a sudden subversion of the hegemonic social bond altogether and does not produce new significations, it is still possible to identify how forces of resistance and transformation are constrained but do not disappear. This means that we do not have to conclude or assume, for example, that *circular economy* is only another ecological modernisation or sustainable development idea tout court. As these traumatic points persist, a final aspect of my analysis is preoccupied with the detection of those forces that overtly introduce disruptive elements, challenge the given hegemonic knowledge, and attempt to bring a real discourse of transformation, what I have called the 'hystericizing' effect, as informed by Lacan's terminology. This hystericizing role has been attributed to the stakeholders in the field such as the EESC who actively lobby the Commission, and to the researchers and activists that populate the unofficial landscape of the EU, such as those observed at the Postgrowth Conference. For example, in the case of the Postgrowth Conference some scientists posed an epistemological question as to what constitutes evidence-based knowledge and challenged the reductionist, rationalised and instrumental use of science as formalised in mathematisation qua modelling and technologisation. Conversely, they contrasted it with a more authentic scientific enquiry that needs to think in complex systems to gain a comprehensive understanding of the real. As a result, issues of 'energy savings', 'efficiency', and 'rebound effect' have been openly called into question by *energy volumes, sufficiency, planetary boundaries*, which might contribute to the creation of new alternative significations and new subjectivities other than the worker-consumer. Further, the case of renewables exposes subjects to the fractures of the discourse more explicitly to the point of questioning the very same mass transition to renewable energy, because of the source intermittency (thus natural capacity), the technicality of the grids (renewable as technology), storage scaling and the final amount of electricity usage available to users. Yet, in the case of *energy efficiency* the traumatic points can also be integrated successfully into signification, and this can occur for instance via the appraisal of the rebound effect as these 'energy savings' are re-invested in the market.

Finally, the case of the circular economy is interesting as it is the case of a defying element from within the EU's policy-making that has managed to shake the foundation of the discourse, although for a limited period. Indeed, unlike *energy efficiency* which can be regarded as the by-product of the University discourse of 'knowledge at work', *circular economy* originally draws on alternative and old circularity philosophies and metaphor of complexity, metabolism, and interrelation. For this reason, this circle metaphor is antithetical to the 'line' metaphor that has characterised our way of exploiting natural resources, and of organising our economy. Perhaps more importantly, this circle metaphor seemed to stand in opposition to those ways of knowing that fail to grasp this system complexity, with its reference to 'metabolism' and 'interrelation'. One of the main arguments was that the circular economy signification ideally forces us to consider the bigger picture of all material and organic flows and how these interact with the surrounding biophysical environment. *Circular economy* could thus potentially entail a common pattern across policy areas, greater symbiosis among policy DGs as well as an expansion of the narrow signification spaces that reflects the linear, reductionist and bureaucratised knowledge of the EU's climate policy. For this reason, *circular economy* was at first resisted to the point of being withdrawn, as it was perceived as a disruptive element in signification, perhaps because it pointed to ideally rethinking the economy as a cycle. This subversive idea in turn can potentially introduce an idea of limit, pace, size to the business-as-usual logic, and to the way 'knowledge' is produced and exchanged within the EU, for example by modelling. Yet, eventually *circular economy* has been accepted into the EU's policy-making and climate action more specifically – that is to say, the climate action + energy transition nexus – but in a co-opted version which has entailed a loss of the traumatic potential that the 'circularity' metaphor brought with it.

Nevertheless, even within this co-opted framework, some subjects such as the EESC were able to keep the hystericizing element alive and visible, by retaining a more transformational understanding of circular economy and pushing its signification beyond its most limited interpretation, which is 'recycling'. Indeed, through these subjects I have illustrated that recycling does not mean that less material is produced in the first place, nor that the amount 'recycled' is the actual material that is re-used, and which re-enters the productive cycle. This is due to the way in which the recycling rate knowledge is produced, as this equates with pre-sorting waste collection, rather than with the secondary raw materials that re-enter the production cycle. Moreover, regardless of how it is measured, *circular economy* qua recycling did not automatically depict a closed loop and circularity: this means that to make a product recyclable, this product must be made recyclable in the first place by means of eco-design, yet even very high recycling rates are compatible with a linear economic model. Therefore, these hysteric subjects are pushing signification beyond the periphery of recycling towards the core of eco-design perhaps because this is what makes the regenerative component possible. Furthermore, they are at the same time challenging the narrow signification of eco-design as energy efficiency, whose signification expansion in the latest Circular Economy Action Plan (2020) is perhaps the most positive example of disturbance of the hegemonic discourse. Finally, more circularity does not necessarily mean an economy that goes at a slower pace, that flows slower than a linear one, or that respects the time necessary for 'regeneration' within an ecosystem. Thus, hysteric subjects such as the EESC 'disturb' signification against a circle that is ever expanding in size and becoming more geographically dispersed and delocalised, and this ideally contributes to *circular economy* being a real discourse of change. At the same time, these 'hysterics' can function within a dominant University/Capitalist social link as scientists, as academics, as activists but their efficacy is affected by the law of that dominant social bond. In fact, in my examples every traumatic encounter with the lack as disruptive force was resisted, minimised, and downplayed. This aspect was shown with events such as the Postgrowth Conference insofar as its relevance in the policy-making was downplayed and minimised by the EU Commission, even if it was organised by EU parliamentary groups and involved the participation of the Commission. Similarly, it was demonstrated by the first withdrawal and then co-optation of the circular economy project.

Of status quo and change in environmental politics and beyond

In summary, in this book I have developed a framework for understanding the EU's climate action in the period 'in-between strategies' against the background of what constitutes a transformative potential and how to investigate that. To accomplish this, I have made Lacan's theoretical framework operational in the empirical sense and developed and implemented a Lacanian discourse analysis which enabled me to emphasise the fractures in the discourse and observe how these are dealt with by the speaking subjects. At the same time, these book's insights are broader in scope for the way in which they discuss agency, power, knowledge, and change, thus this conclusive section makes the case for its relevance in environmental politics and IR more generally.

In general, my analysis has revolved around the politics of knowledge production in a way that pays attention to the disavowal of any real authority of the discourse even when this manifests through objects that are apparently desirable and not particularly contentious. Mitigation knowledge tools such as energy efficiency, renewables and circular economy are in principle desirable, yet I have demonstrated that the knowledge underpinning these can reveal its political and performative nature which is often disavowed under a flat and apparent neutrality and objectivity. Notably, the exposure of the Real of the discourse and its traumatic and liberating potential shed a light on the mechanisms governing status quo and change insofar as we can appreciate how the impossibility of the discourse triggers and fuels desire; this ultimately leads subjects to cope with the effects and deficiencies of the discourse, by both paradoxically enjoying a discourse that repeatedly fails them (*jouissance*) and by constructing fantasies that promise to eliminate the lack and achieve the desired object. This investigation on knowledge and the fantasies and enjoyments that animate these can be expanded and applied to other climate plans and initiatives that have recently reappeared in the climate action imagery – within and beyond the EU – such as CCS technologies. The case of CCS is interesting because its policy project had fallen flat due to controversies regarding its social acceptance, technology feasibility and potential geophysical consequences; however, it has resurfaced again in the political and scientific arena as one of the levers of mitigation and hailed as a necessary solution for energy intensive industries although it is still not quite an available technology, especially on a large scale. The impossibility of the CCS discourse – in the Lacanian sense – is also revealed when CCS is accompanied by a newer and softer variation, named Carbon Capture and Utilisation (CCU), to those processes by which CO_2 is captured and converted and thus re-used and re-emitted into a new product, assumingly with reduced mitigating effects.

Moreover, in Chapter 2 I illustrated that a few evolutions have characterised the EU's climate action since this research was conducted, and that the phase of climate action under observation in this book constitutes only a snapshot of a wider and longer process. The EU Green Deal (COM (2019) 640 final) has become the overarching framework in which environmental – including climate – policies are formulated. The launch of the Green Deal and the subsequent work for its implementation represent a fertile ground for reflection on power, knowledge(s) and agency due to the historical moment in which it emerged. More specifically, after its publication as a Communication in December 2019, a period of stalemate inevitably followed due to the outbreak of the COVID-19 pandemic of which – from a mere theoretical speculation – a symptomatic reading can be provided. It is now common knowledge that in 2020 the origin and spread of the virus – which in itself urged us to reflect on the constant intrusion of humans into wildlife habitats – had spurred ethical dilemmas over saving lives versus saving the economy, and that this dilemma was handled with different social and economic lockdowns, to the temporary benefit of the climate and the environment (Le Quéré et al., 2020). Thus, the pandemic experience could have constituted the ultimate 'hystericizing' element carrying a truly liberating and traumatic force for its socio-economic consequences, capable of opening a painful and irreversible fracture in the discourse and opening the conditions for a truly transformative discourse, by means of the Lacanian Analyst's discourse. Indeed, in this discourse *objet petit a* interrogates the subject in their split and exposes the hysterical structure of the discourse to make them identify the Masters Signifiers that form the alienating identification: this is a painful and disruptive process often dominated by anxiety and meaninglessness (Bracher, 1993) which allows for the resurfacing of the left out and repressed. The final goal of this process is to eventually expose and traverse the underlying 'fantasy' – namely recognise the fictitiousness of the socio-symbolic order and understanding that those (unconscious) fantasies that have been driving one's desire are in fact relative – to eventually arrive to a new Master Signifier and a new subject (ibid., 1993). Yet, the EU's climate policy activity within the Green Deal that characterised the post pandemic is far from having traversed the fantasy and seems to bear little of the traumatic force that the pandemic experience has represented, although it is nonetheless a field of struggle of knowledge(s) and paradoxes. For example, while the EU Climate Law and the increased targets of the Fit for 55 package are in line with the EU's subjectivity as a regulatory entity and do not seem to break with the modus operandi of the hegemonic University discourse of knowledge and Capitalist discourse of *jouissance*, the 'environmentalisation' and cross-sectoral ambition of the Green Deal beyond the climate-energy nexus, perpetrated by the seemingly more progressive stances of Ursula Von Der Leyen's Commission, point to less reductionist knowledge compared to the examples provided in this book, and can still be interpreted as positive acts of knowledge hystericization. However, the watering down of the Nature Restoration Law (EU Parliament, 2024) and the scrapping of the proposed new Regulation on pesticides due to the strong opposition from farmers and pesticides industry – notably in the run-up to the 2024 Parliamentary elections (Weise and Guillot, 2024) – would point once again to a deadlock of the Green Deal and a drastic reduction of its real transformative potential. Within this framework, the fantasies and enjoyments that animate, for instance, both these protests and the reactions to these can constitute the entry point for further symptomatic critiques. More to the point, the EU Commission's handling of the farmers' protests – which ended with the withdrawal of the proposed legislation – is emblematic. On the one hand the farmers' subjectivity is constituted through a discourse of knowledge that sees them as perpetrators of intensive farming and agricultural practices that are detrimental to the soil, to biodiversity and to the climate and which needs to be tightly regulated; on the other hand the Free Trade Agreements that the EU proudly continues to strike with the rest of the world (European Commission, 2024) – the by-product of the Capitalist discourse of enjoyment – expose these subjects to a competition that cannot be questioned. This in turn creates a fertile ground for social discontent that is in turn exploited by nationalist governments and becomes an ideological battleground within the domestic politics of the Member States.

Finally, this research idea for this book originated from the language turn in IR insofar as it shared the centrality of language in the constitution of the social and political world – in which climate policies are situated – and enabled a non-reductionist investigation of the socio-political dimension of climate change, which takes into account the complexity of actors and institutions that populate environmental politics and governance in general, such as decision-makers, stakeholders, scientists and academics and citizens. Consistently, by focusing on the 'discourse' I illustrated that a given policy outcome, such as the Energy Efficiency Directive (2012/27/EU, (EU) 2018/2002), or the Renewable Energy Directive (2009/28/EC. 2018/2001/EU), or the Circular Economy Action Plan (COM (2015) 614 final, COM/2020/98 final), can be considered as a discursive and material implication of a given discourse. This has situated my analysis in line with previous

Foucauldian approaches that stressed the socio-cultural meaning structures, as well as the discursive attributes of power-knowledge relationships, (Feindt and Oels, 2005; Hajer, 1995; Hajer and Versteeg, 2005; Litfin, 1994; Oels, 2005) and this aspect emerged in my analysis of the commanded climate mitigation knowledge, of which I however emphasised an important element of disavowal concerning its real authority.

Yet, this work departed from existing Foucauldian discursive approaches in two main respects. At theoretical level, it reconfigured the classical IR structure versus agency debate and retained a more resistant subject, to the extent that it pointed to a dialectical relationship between discourse and subject by introducing a constitutive (ontological) lack – in which the socio-linguistic order can never complete the subjects – whereby subjects emerge as produced by language but as a surplus of sense. Nevertheless, concluding that this subject is 'more resistant' must not mislead us into believing that the agent is unconstrained and more or less able to act influencing the processes and structures. In fact, the analysis of the potentially transformative forces in the field revealed how these are at the moment resisted, minimised, and downplayed and how these hysterics can still function within the dominant status quo and question the hegemonic social link, yet their efficacy is affected by the law of the dominant social bond, Lacan's University and Capitalist discourses. At the same time, by virtue of a more active subject, whose role can be understood by looking at how this constitutive lack is handled, we can better understand factors of stability, rupture, and change, and notably distinguish between an apparent fictitious change and real 'change', which would be difficult to explain by referring to a subject that is purely 'structural' and purely subjugated to power. In this respect this work adds up to existing contributions applying a psychoanalytical filter to global political issues such as development (Kapoor, 2020), nationalism (Mandelbaum, 2022), security and justice (Zevnik, 2017) and terrorism (Solomon, 2015).

At methodological level, I demonstrated how Lacan's theory of discourse can be made operational in the empirical sense and I illustrated how a Lacanian discourse analysis can provide a useful empirical method in IR that allows us to go beyond an understanding of 'discourse' as competing, full, achieved 'storylines', narratives, or framings (Hajer, 1995; Leipold et al., 2019, 457). For instance, in my analysis *energy efficiency, renewables,* or *circular economy* are not regarded as 'discourses' qua narratives but as the inevitable derivatives of the discourse qua social bond, that is the by-product of that socio-linguistic structure that makes it possible to establish, define and maintain our societal relationships. Further, an emphasis on the enunciating act as the moment when the socio-symbolic and the excess-leftover of sense qua Lacanian subject emerge made it possible to expose the fractures of the discourse, that is to demonstrate that full meaning is always deferred across the signifying chain, thus revealing the partial and never-achieved character of the signifier. This way, if we wanted to keep the metaphor of the 'storyline', it would be more accurate to assert that each spoken signifier at different times of the enunciation gives origin to different and always partial storylines of 'mitigation', of 'energy efficiency', of 'circular economy' which are not 'competing' but coexist – together with their inconsistencies and their associated fantasies and forms of enjoyment – in giving meaning to the signifier, as a result of the surplus-loss of signification produced by the speaking subject during the enunciation. This argument developed a conceptualisation of the discourse that is more radical compared to that of Hajer and Versteeg (2005), who maintained that a focus on the discourse helps understand how different actors actively attempt to influence the definition of an issue by imposing a given frame and by drawing on discursive categories in creating a coalition (Hajer, 1995). From that perspective misunderstandings can be functional to the pursuit of different interests in the political game and although actors assume mutual understanding and use storylines to debate 'nature' or 'climate change' in shared terms, this does not mean that they understand each other.

By contrast, while in a Lacanian approach subjects do resort to a presupposed battery of signifiers which is thus assumed in shared terms, and while misunderstanding and ambiguity are still key, enunciation brings with it unexpected effects, and for this reason each signifier takes different forms and shapes in the enunciation, which is ultimately the Lacanian subject produced by language as surplus of sense. Hence, rather than a deliberate and active attempt to build a 'narrative' or a 'storyline' of energy efficiency, of renewables or circular economy, these inconsistencies and continuous misunderstandings are the *unintended* effects of the signifying chain on the speaking subjects and their copying mechanisms in dealing with a discourse that is ultimately impossible. A linguistic Lacanian discourse analysis interpreted in conjunction with his theory of the four discourses allows us to open and disrupt a seemingly closed and consistent (hegemonic) discourse and expose the Real of the discourse, that is how the excess of meaning produced by the subjects – evident in gaps, weaknesses, and contradictions – is handled in terms of fantasies and enjoyments. Ultimately, it provides one possible way to think about and investigate agency and change in their complexity in the field of contemporary IR and makes it possible to assess whether and how historicised dominant knowledge and its authority are embraced, disavowed, questioned, and potentially subverted.

Bibliography

Adler, E. (1997) "Seizing the Middle Ground: Constructivism in World Politics." *European Journal of International Relations*, 3(3), 319–363.

Adler-Nissen, E., and Pouliot, V. (2011) *International Practices*. Cambridge: Cambridge University Press.

Audet, R. (2016) "Transition as Discourse." *International Journal of Sustainable Development*, 19 (4), 365–382.

Beck, U. (1995) *Ecological Politics in an Age of Risk*. Cambridge: Polity Press.

Bennett, J. (2010) *Vibrant Matter: A Political Ecology of Things*. Durham, NC and London: Duke University Press.

Bettini, G. (2019) "And Yet it Moves! (Climate) Migration as a Symptom in the Anthropocene." *Mobilities*, 14 (3), 336–350.

Boni, L. (2014) "Formalisation and Context: Some Elements of a Materialist Reading of Lacan's 'Four Discourses'." In I. Parker and D. Pavón-Cuéllar (eds), *Lacan, Discourse, Event: New Psychoanalytic Approaches to Textual Indeterminacy*. London and New York: Routledge, 128–139.

Bouckaert, R., and Dupont, C. (2022) "Turning to Algeria to Replace Russian Gas: A False Solution." GOVTRAN Policy Brief 2/2022. www.brussels-school.be/sites/default/files/documents/Turning%20to%20Algeria%20to%20replace%20Russian%20gas.pdf (last accessed May 2024).

Bracher, M. (1993) *Lacan, Discourse and Social Change: A Psychoanalytic Cultural Criticism*. Ithaca and London: Cornell University Press.

Bracher, M., and Alcorn, M. (1994) *Lacanian Theory of Discourse*. New York: New York University Press.

Bretherton, C., and Vogler, J. (2006) *The European Union as a Global Actor*. 2nd ed. London: Routledge.

Burgess, J. P. (2017) "The *Real* at the Origin of Sovereignty." *Political Psychology*, 38 (4), 653–668.

Butler, J. (1997) *The Psychic Life of Power: Theories in Subjection*. Stanford: Stanford University Press.

Campbell, D. (1992) *In Writing Security: United States Foreign Policy and the Politics of Identity*. Ann Arbor: University of Minnesota Press.

Campbell, K. (2016) "Political Encounters: Feminism and Lacanian Psychoanalysis." In S. Tomšič and A. Zevnik (eds), *Jacques Lacan: Between Psychoanalysis and Politics*. London: Routledge, 233–252.

Chomsky, N. (1981) *Lectures on Government and Binding: The Pisa Lectures*. Holland: Foris Publications (Reprint 7th ed.). Berlin and New York: Mouton de Gruyter, 1993.

Ciplet, D., Roberts, T. J., and Khan, M. R. (2015) *Power in a Warming World*. Cambridge: MIT Press.

Connolly, W. E. (1974) *The Terms of Political Discourse*. Lexington, MA: D.C. Heath and Co.

Coole, D. (2013) "Agentic Capacities and Capacious Historical Materialism: Thinking With New Materialisms in the Political Sciences." *Millennium: Journal of International Studies*, 41 (3), 451–469.

Copjec, J. (1994) *Read My Desire: Lacan Against the Historicists*. London: Verso.

Curtin, D., and Manucharyan, T. (2015) "Legal Acts and Hierarchy of Norms in EU Law." In A. Arnull and D. Chalmers (eds), *The Oxford Handbook of European Union Law*. Oxford: Oxford University Press, 103–125.

De Saussure, F. (1959) *Course in General Linguistics*. New York: Philosophical Library.

Der Derian, J. (1987) *On Diplomacy*. Oxford: Blackwell.

Dimitrov, R. S. (2016) "The Paris Agreement on Climate Change: Behind Closed Doors." *Global Environmental Politics*, 16 (3), 1–11.

Doty, R. (1993) "Foreign Policy as Social Construction: A Post-Positivist Analysis of US Counterinsurgency Policy in the Philippines." *International Studies Quarterly*, 37 (3), 297–320.

Dreyfus, H. L., and Rainbow, P. (1983) *Foucault: Beyond Structuralism and Hermeneutics*. Chicago, IL: University of Chicago Press.

Drieschova, A. (2017) "Peirce's Semeiotics: A Methodology for Bridging the Material-Ideational Divide in IR Scholarship." *International Theory*, 9 (1), 33–66.

Dryzek, J. S. (1997) *The Politics of the Earth: Environmental Discourses*. New York: Oxford University Press.

Dunn, K., and Neumann, I. (2016) *Undertaking Discourse Analysis for Social Research*. Ann Arbor: University of Michigan Press.

Dupont, C. (2016) *Climate Policy Integration into EU Energy Policy: Progress and Prospects*. London: Routledge.

Dupont, C., Moore, B., Boasson, E. L., Gravey, V., Jordan, A., Kivimaa, P., Kulovesi, K., Kuzemko, C., Oberthür, S., Panchuk, D., and Rosamond, J. (2023) "Three Decades of EU Climate Policy: Racing Toward Climate Neutrality?" *WIREs Climate Change*, 15 (1), e863.

Eckert, S. G. (2021) "The European Green Deal and the EU's Regulatory Power in Times of Crisis." *Journal of Common Market Studies*, 59 (S1), 81–91.

EEA. (2023) "EU Achieves 20–20–20 Climate Targets, 55 % Emissions Cut by 2030 Reachable With More Efforts and Policies." www.eea.europa.eu/highlights/eu-achieves-20-20-20#:~:text=20%2D20%2D20%20targets%20achieved,of%20the%2020%20%25%20 reduction%20target (last accessed May 2024).

EESC. (2013) "Towards More Sustainable Consumption: Industrial Product Lifetimes and Restoring Trust Through Consumer Information." CCMI/112-EESC-2013–1904. www.eesc.europa.eu/en/our-work/opinions-information-reports/opinions/towards-more-sustainable-consumption-industrial-product-lifetimes-and-restoring-trust-through-consumer-information#downloads (last accessed May 2024).

Ellen MacArthur Foundation. (2019) "What Is a Circular Economy." www.ellenmacarthur-foundation.org/circular-economy/concept (last accessed May 2024).

Ellen MacArthur Foundation, McKinsey Centre for Business and Environment and Stiftungsfonds für Umweltökonomie und Nachhaltigkeit (SUN). (2015) "Growth Within: A Circular Economy Vision for a Competitive Europe." www.mckinsey.com/capabilities/sustainability/our-insights/growth-within-a-circular-economy-vision-for-a-competitive-europe (last accessed May 2024).

Epstein, C. (2008) *The Power of Words in International Relations: Birth of an Anti-Whaling Discourse*. Cambridge: Cambridge University Press.

Epstein, C. (2011) "Who Speaks? Discourse, the Subject and the Study of Identity in International Politics." *European Journal of International Relations*, 17 (2), 327–350.

EU. (2009a) "Renewable Energy Directive." Directive 2009/28/EC. https://eur-lex.europa.eu/legal-content/EN/ALL/?uri=CELEX%3A32009L0028 (last accessed May 2024).

EU. (2009b) "Renewable Energy Directive." Directive 2018/2001. https://eur-lex.europa.eu/legal-content/en/TXT/?uri=CELEX%3A32018L2001 (last accessed May 2024).

EU. (2012) "Energy Efficiency Directive." Directive 2012/27/EU. https://eur-lex.europa.eu/legal-content/EN/TXT/?qid=1399375464230&uri=CELEX:32012L0027 (last accessed May 2024).

EU. (2018a) "Energy Efficiency Directive." Directive EU 2018/2002. https://eur-lex.europa.eu/legal-content/EN/TXT/?uri=celex%3A32018L2002 (last accessed May 2024).

EU. (2018b) "Regulation (EU) 2018/1999 of the European Parliament and of the Council of 11 December 2018 on the Governance of the Energy Union and Climate Action." https://eur-lex.europa.eu/legal-content/EN/TXT/?uri=uriserv:OJ.L_.2018.328.01.0001.01.ENG&toc=OJ:L:2018:328:FULL (last accessed May 2024).

EU. (2021) "Regulation (EU) 2021/1119 of the European Parliament and of the Council of 30 June 2021 Establishing the Framework for Achieving Climate Neutrality and Amending Regulations (EC) No 401/2009 and (EU) 2018/1999 ('European Climate Law')." https://eur-lex.europa.eu/legal-content/EN/TXT/?uri=CELEX:32021R1119 (last accessed May 2024).

EU. (2023) "Regulation (EU) 2023/851 of the European Parliament and of the Council of 19 April 2023 Amending Regulation (EU) 2019/631 as Regards Strengthening the CO_2 Emission Performance Standards for New Passenger Cars and New Light Commercial Vehicles in Line With the Union's Increased Climate Ambition." https://eur-lex.europa.eu/legal-content/EN/TXT/?uri=celex:32023R0851 (last accessed May 2024).

European Commission. (2014) "Communication: Towards a Circular Economy: A Zero Waste Programme for Europe." COM (2014) 398 Final. Brussels: European Commission. https://eur-lex.europa.eu/legal-content/EN/TXT/?uri=CELEX%3A52014DC0398 (last accessed May 2024).

European Commission. (2015) "Communication: Closing the Loop – An EU Action Plan for the Circular Economy." COM (2015) 614 Final. Brussels: European Commission. https://eur-lex.europa.eu/legal-content/EN/TXT/?uri=CELEX:52015DC0614 (last accessed May 2024).

European Commission. (2016) "Communication Eco-Design Working Plan 2016–2019." COM (2016) 0773 Final. Brussels: European Commission. https://eur-lex.europa.eu/legal-content/EN/TXT/?qid=1591200883158&uri=CELEX:52016DC0773 (last accessed May 2024).

European Commission. (2018a) "A Clean Planet for All – A European Strategic Long-Term Vision for a Prosperous, Modern, Competitive and Climate Neutral Economy." COM (2018) 773 Final. Brussels: European Commission. https://eur-lex.europa.eu/legal-content/en/TXT/?uri=CELEX%3A52018DC0773 (last accessed May 2024).

European Commission. (2018b) "A Clean Planet for All – A European Strategic Long-Term Vision for a Prosperous, Modern, Competitive and Climate Neutral Economy." COM (2018) 773 Final, in Depth Analysis Accompanying the Communication. https://climate.ec.europa.eu/system/files/201811/com_2018_733_analysis_in_support_en.pdf (last accessed May 2024).

European Commission. (2019) "Communication: The European Green Deal." COM (2019) 640 Final. Brussels: European Commission. https://eur-lex.europa.eu/legal-content/EN/TXT/?qid=1591201354409&uri=CELEX:52019DC0640 (last accessed May 2024).

European Commission. (2020a) "Circular Economy Action Plan." https://ec.europa.eu/environment/circular-economy/first_circular_economy_action_plan.html (last accessed May 2024).

European Commission. (2020b) "Communication: A New Circular Economy Action Plan for a Cleaner and More Competitive Europe." COM/2020/98 Final. Brussels: European Commission. https://eur-lex.europa.eu/legal-content/EN/TXT/?qid=1583933814386&uri=COM:2020:98:FIN (last accessed May 2024).

European Commission. (2022) "Communication: REPowerEU Plan." COM(2022) 230 Final. https://eur-lex.europa.eu/legal-content/EN/TXT/?uri=COM%3A2022%3A230%3AFIN&qid=1653033742483 (last accessed May 2024).

European Commission, DG Energy. (2019) "Clean Energy for All Europeans." https://op.europa.eu/en/publication-detail/-/publication/b4e46873-7528-11e9-9f05–01aa75ed71a1/

language-en?WT.mc_id=Searchresult&WT.ria_c=null&WT.ria_f=3608&WT.ria_ev=search (last accessed May 2024).

European Commission, DG Research and Innovation. (2018) *Final Report of the High-Level Panel of the European Decarbonisation Pathways Initiative*. Brussels: European Commission. https://op.europa.eu/en/publication-detail/-/publication/226dea40-04d3-11e9-adde-01aa75ed71a1 (last accessed May 2024).

European Council/Council of the European Union. (2022) "Fit for 55." www.consilium.europa.eu/en/policies/green-deal/fit-for-55-the-eu-plan-for-a-green-transition/ (last accessed May 2024).

European Parliament. (2017) "Motion for a European Parliament Resolution on a Longer Lifetime for Products: Benefits for Consumers and Companies." www.europarl.europa.eu/doceo/document/A-8-2017-0214_EN.html (last accessed May 2024).

European Parliament. (2024) "Nature Restoration: Parliament Adopts Law to Restore 20% of EU's Land and Sea." www.europarl.europa.eu/news/en/press-room/20240223IPR18078/nature-restoration-parliament-adopts-law-to-restore-20-of-eu-s-land-and-sea (last accessed May 2024).

Evans, D. (1996) *An Introductory Dictionary of Lacanian Psychoanalysis*. London: Routledge.

Fairclough, N. (2010) *Critical Discourse Analysis: The Critical Study of Language*. 2nd ed. Longman Applied Linguistics. Harlow: Longman.

Feindt, P. H., and Oels, A. (2005) "Does Discourse Matter? Discourse Analysis in Environmental Policy Making." *Journal of Environmental Policy & Planning*, 7 (3), 161–173.

Feldner, H., and Vighi, F. (2015) *Critical Theory and the Crisis of Contemporary Capitalism*. New York [u.a.]: Bloomsbury Academic.

Fink, B. (1999) "The Master Signifier and the Four Discourses." In D. Nobus (ed), *Key Concepts of Lacanian Psychoanalysis*. New York: Other Press.

Fischer, F. (2003) *Reframing Public Policy: Discursive Politics and Deliberative Practices*. Oxford: Oxford University Press.

Fletcher, R., and Rammelt, C. (2017) "Decoupling: A Key Fantasy of the Post-2015 Sustainable Development Agenda." *Globalizations*, 14 (3), 450–467.

Foucault, M. (1972) *The Archaeology of Knowledge*. London: Pantheon.

Foucault, M. (1977) *Discipline and Punish*. New York: Pantheon Books.

Foucault, M. (1979) *The History of Sexuality*. London: Allen Lane.

Foucault, M. (1980) "Truth and Power." In M. Foucault and C. Gordon (eds), *Power/Knowledge*. 1st ed. New York: Pantheon Books.

Gheuens, J., and Oberthür, S. (2021) "EU Climate and Energy Policy: How Myopic Is It?" *Politics and Governance*, 9 (3), 337–347.

Gravey, V., and Jordan, A. (2021) "New Policy Dynamics in More Uncertain Times?" In V. Gravey and A. Jordan (eds), *Environmental Policy in the EU*. Abingdon: Routledge, 334–354.

Griffin, P. (2009) *Gendering the World Bank: Neoliberalism and the Gendered Foundations of Global Governance*. Basingstoke: Palgrave Macmillan.

Gupta, J., and Grubb, M. (2000) *Climate Change and European Leadership*. 1st ed. Dordrecht: Springer.

Habermas, J. (1996) *Between Facts and Norms: Contributions to a Discourse Theory of Law and Democracy*. Cambridge, MA: MIT Press.

Hajer, M. (1995) *The Politics of Environmental Discourse*. 1st ed. Oxford: Clarendon Press.

Hajer, M., and Versteeg, W. (2005) "A Decade of Discourse Analysis of Environmental Politics: Achievements, Challenges, Perspectives." *Journal of Environmental Policy & Planning*, 7 (3), 175–184.

Hajer, M., and Wagenaar, H. (eds). (2003) *Deliberative Policy Analysis: Understanding Governance in the Network Society*. Cambridge: Cambridge University Press.

Hansen, L. (2006) *Security as Practice: Discourse Analysis and the Bosnian War*. London: Routledge.

Haraway, D. (1991) *Simians, Cyborgs, and Women: The Reinvention of Nature*. New York: Routledge.

Herschinger, E. (2011) *Constructing Global Enemies: Hegemony and Identity in International Discourses on Terrorism and Drug Prohibition*. London: Routledge.

Hook, D. (2013) "Tracking the Lacanian Unconscious in Language." *Psychodynamic Practice*, 19 (1), 38–54.

Hughes, H. (2015) "Bourdieu and the IPCC's Symbolic Power." *Global Environmental Politics*, 15 (4), 85–104.

Hulme, M. (2008) "The Conquering of Climate: Discourses of Fear and Their Dissolution." *The Geographical Journal*, 174 (1), 5–16.

Hurka, S., Haag, M., and Kaplaner, C. (2021) "Policy Complexity in the European Union, 1993-Today: Introducing the EUPLEX Dataset." *Journal of European Public Policy*, 29 (9), 1512–1527.

IEA. (2023) "Understanding Methane Emissions." www.iea.org/reports/global-methane-tracker-2023/understanding-methane-emissions (last accessed May 2024).

IPCC (2007): *Climate Change 2007: Synthesis Report. Contribution of Working Groups I, II and III to the Fourth Assessment Report of the Intergovernmental Panel on Climate Change* [Core Writing Team, R. K. Pachauri and A. Reisinger (eds.)]. Geneva: IPCC, 104 pp.

IPCC. (2014) *Climate Change 2014: Synthesis Report. Contribution of Working Groups I, II and III to the Fifth Assessment Report of the Intergovernmental Panel on Climate Change* [Core Writing Team, R. K. Pachauri, and L. A. Meyer (eds.)]. Geneva: IPCC, 151.

IPCC. (2018) *Global Warming of 1.5°C. An IPCC Special Report on the Impacts of Global Warming of 1.5°C Above Pre-Industrial Levels and Related Global Greenhouse Gas Emission Pathways, in the Context of Strengthening the Global Response to the Threat of Climate Change, Sustainable Development, and Efforts to Eradicate Poverty* [V. Masson-Delmotte, P. Zhai, H. O. Pörtner, D. Roberts, J. Skea, P. R. Shukla, A. Pirani, W. Moufouma-Okia, C. Péan, R. Pidcock, S. Connors, J. B. R. Matthews, Y. Chen, X. Zhou, M. I. Gomis, E. Lonnoy, T. Maycock, M. Tignor, T. Waterfield (eds.)]. In Press.

IPCC. (2023) "Summary for Policymakers." In Core Writing Team, H. Lee, and J. Romero (eds), *Climate Change 2023: Synthesis Report. Contribution of Working Groups I, II and III to the Sixth Assessment Report of the Intergovernmental Panel on Climate Change*. Geneva: IPCC, 1–34. DOI: 10.59327/IPCC/AR6-9789291691647.001.

Jordan, A., Huitema, D., van Asselt, H., Rayner, T., and Berkhout, F. (2010) *Climate Change Policy in the European Union. An Introduction*. Cambridge: Cambridge University Press.

Kapoor, I. (2020) *Confronting Desire: Psychoanalysis and International Development*. Ithaca, NY: Cornell University Press.

Keller, R. (2012) *Doing Discourse Research: An Introduction for Social Scientists*. London: Sage.

Keller, R., Hornidge, A. K., and Schünemann, W. (2018) *The Sociology of Knowledge Approach to Discourse*. London: Routledge.

Klepec, P. (2016) "On the Mastery in the Four 'Discourses'." In S. Tomšič Samo and A. Zevnik (eds), *Jacques Lacan: Between Psychoanalysis and Politics*. London: Routledge, 115–130.

Koren, D. (2014) "Agonistic Discourses, Analytic Act, Subjective Event." In I. Parker and D. Pavón-Cuéllar (eds), *Lacan, Discourse, Event: New Psychoanalytic Approaches to Textual Indeterminacy*. London and New York: Routledge, 247–256.

Korhonen, J., Honkasalo, A., and Seppälä, J. (2018) "Circular Economy: The Concept and Its Limitations." *Ecological Economics*, 143, 37–46.

Kulovesi, K., and Oberthür, S. (2020) "Assessing the EU's 2030 Climate and Energy Policy Framework: Incremental Change Toward Radical Transformation?" *Review of European, Comparative and International Environmental Law*, 29 (2), 151–166.

Lacan, J. (1964) *Le Séminaire. Livre XI. Les Quatre Concepts Fondamentaux de la Psychanalyse*. Paris: Seuil.

Lacan, J. (1968–1969/2006) *Le Séminaire: Livre XVI. D'un autre à l'Autre*. Paris: Seuil.

Lacan, J. (1969–1970/1991) *Le Séminaire. Livre XVII. L'Envers de la Psychanalyse*. Paris: Seuil.

Lacan, J. (1972) "Du discours psychanalitique/Del discorso psicoanalitico." In *Lacan in Italia, 1953–1978*. Milano: La Salmandra, available online.

Lacan, J. (1973) "Excursus/Excursus." In *Lacan in Italia, 1953–1978*. Milano: La Salmandra, available online.

Lacan, J. (1988) *The Seminar of Jacques Lacan, Book I: Freud's Papers on Technique, 1953–1954* (J. Forrester, Trans.). Cambridge: Cambridge University Press.

Lacan, J. (2007) *The Seminar, Book XVII: The Other Side of Psychoanalysis*. New York and London: W.W. Norton.

Lacan, J. (2013) *Le Séminaire VI, Le désir et son interprétation, 1958–1959*. Paris: Seuil.

Lacan, J., Miller, J. A., Copjec, J., Hollier, D., Krauss, R., Michelson, A., and Mehlman, J. (1990) *Television*. New York: W.W. Norton.

Lacan, J., Miller, J. A., and Sheridan, A. (1998) *The Four Fundamental Concepts of Psycho-Analysis*. New York: Norton and Company.

Laclau, E. (1996) *Emancipation(s)*. London: Verso.

Laclau, E., and Mouffe, C. (1985) *Hegemony and Socialist Strategy: Towards a Radical Democratic Politics*. 2nd ed. London and New York: Verso.

Le Quéré, C., Jackson, R. B., Jones, M. W., Smith, A. J., Abernethy, S., Andrew, R. M., De-Gol, A. J., Willis, D. R., Shan, Y., Canadell, J. G., and Friedlingstein, P. (2020) "Temporary Reduction in Daily Global CO_2 Emissions During the COVID-19 Forced Confinement." *Nature Climate Change*, 10 (7), 647–653.

Leipold, S., Feindt, P. H., Winkel, G., and Keller, R. (2019) "Discourse Analysis of Environmental Policy Revisited: Traditions, Trends, Perspectives." *Journal of Environmental Policy & Planning*, 21 (5), 445–463.

Litfin, K. (1994) *Ozone Discourses*. 1st ed. New York: Columbia University Press.

Lundborg, T., and Vaughan-Williams, N. (2015) "New Materialisms, Discourse Analysis, International Relations: A Radical Intertextual Approach." *Review of International Studies*, 41, 3–25.

Mandelbaum, M. (2022) "The Repetitions of Nationalism: Ontology, Fantasy and Jouissance." In P. de Orellana and N. Michelsen (eds), *Global Nationalism: Ideas, Movements and Dynamics in the Twenty-First Century*. London: World Scientific.

Milliken, J. (1999) "The Study of Discourse in International Relations." *European Journal of International Relations*, 5 (2), 225–254.

Morrison, D. (2003) "New Labour and the Ideological Fantasy of the Good Citizen." *Journal for the Psychoanalysis of Culture and Society*, 8 (2), 273–278.

Neill, C. (2013) "Breaking the Text: An Introduction to Lacanian Discourse Analysis." *Theory & Psychology*, 23 (3), 334–350.

Neumann, I. B. (2002) "Returning Practice to the Linguistic Turn: The Case of Diplomacy." *Millennium: Journal of International Studies*, 31 (3), 627–651.

Newman, S. (2004) "The Place of Power in Political Discourse." *International Political Science Review*, 25 (2), 139–157.

Oberthür, S. (2009) "The Role of the EU in Global Environmental and Climate Governance." In M. Telò (ed), *The European Union and Global Governance*. 1st ed. London: Routledge, 192–209.

Oberthür, S., and Dupont, C. (2021) "The European Union's International Climate Leadership: Towards a Grand Climate Strategy?" *Journal of European Public Policy*, 28 (7), 1095–1114.

Oberthür, S., and Kelly, C. R. (2008) "EU Leadership in International Climate Policy: Achievements and Challenges." *International Spectator*, 43 (3), 35–50.

Oberthür, S., and Pallemaerts, M. (eds). (2010) *The New Climate Policies of the European Union: Internal Legislation and Climate Diplomacy* (IES Publication Series). Brussels: VUB Press.

Oberthür, S., and von Homeyer, I. (2023) "From Emissions Trading to the European Green Deal: The Evolution of the Climate Policy Mix and Climate Policy Integration in the EU." *Journal of European Public Policy*, 30 (3), 445–468.

Oels, A. (2005) "Rendering Climate Change Governable: From Biopower to Advanced Liberal Government?" *Journal of Environmental Policy & Planning*, 7 (3), 185–207.

Parker, I. (2005) "Lacanian Discourse Analysis in Psychology." *Theory & Psychology*, 15 (2), 163–182.

Parker, I. (2010) "Psychosocial Studies: Lacanian Discourse Analysis Negotiating Interview Text." *Psychoanalysis, Culture & Society*, 15 (2), 156–172.

Parker, I. (2014) "Lacanian Discourse Analysis: Seven Elements." In I. Parker and D. Pavón-Cuéllar (eds), *Lacan, Discourse, Event: New Psychoanalytic Approaches to Textual Indeterminacy*. London and New York: Routledge, 38–51.

Pavón-Cuéllar, D. (2014) "The Enunciating Act and the Problem of the Real in Lacanian Discourse Analysis." In I. Parker and D. Pavón-Cuéllar (eds), *Lacan, Discourse, Event: New Psychoanalytic Approaches to Textual Indeterminacy*. London and New York: Routledge, 66–76.

Pavon Cuellar, D., Carlo, D., and Parker, I. (2010) *From the Conscious Interior to an Exterior Unconscious: Lacan, Discourse Analysis and Social Psychology*. London: Karnac.

Pelkmans, J. (2016) "Why the Single Market Remains the EU's Core Business." *West European Politics*, 39 (5), 1095–1113.

Rasiński, L. (2011) "The Idea of Discourse in Poststructuralism: Derrida, Lacan and Foucault." *Teraźniejszość Człowiek Edukacja*: *A Quarterly of Social and Educational Ideas*, 1, 7–22.

Richardson, T., and Sharp, L. (2001) "Reflections on Foucauldian Discourse Analysis in Planning and Environmental Policy Research." *Journal of Environmental Policy & Planning*, 3, 193–209.

Roe, E. M. (1994) *Narrative Policy Analysis: Theory and Practice*. Durham, NC and London: Duke University Press.

Schiffrin, D., and Tannen, D. (2001) *The Handbook of Discourse Analysis*. Oxford: Blackwell.

Shapiro, M. (1981) *Language and Political Understanding: The Politics of Discursive Practices*. New Haven: Yale University Press.

Shepherd, L. (2013) *Critical Approaches to Security*. London: Routledge, 196–207.

Skjærseth, J. B. (2021) "Towards a European Green Deal: The Evolution of EU Climate and Energy Policy Mixes." *International Environmental Agreements: Politics, Law and Economics*, 21 (1), 25–41.

Skjærseth, J. B., and Wettestad, J. (2010) "The EU Emissions Trading System Revised (Directive 2009/29/EC)." In S. Oberthür and M. Pallemaerts (eds), *The New Climate Policies of the European Union*. 1st ed. Brussels: VUB Press.

Solomon, T. (2015) *The Politics of Subjectivity in American Foreign Policy Discourse*. Ann Arbor: University of Michigan Press.

Stavrakakis, Y. (1999) *Lacan and the Political*. London: Routledge.

Šumič, J. (2016) "Politics and Psychoanalysis in the Times of the Inexistent Other." In S. Tomšič Samo and A. Zevnik (eds), *Jacques Lacan: Between Psychoanalysis and Politics*. London: Routledge, 28–42.

Tolis, V. (2023) "A Method for Change: Lacanian Discourse Analysis: A Glimpse into Climate Policy." *International Studies Quarterly*, 67 (3), sqad059.

Tomsič, S. (2016) "Psychoanalysis, Capitalism, and Critique of Political Economy: Toward a Marxist Lacan." In S. Tomšič and A. Zevnik (eds), *Jacques Lacan: Between Psychoanalysis and Politics*. London: Routledge, 146–163.

Tomšič, S., and Zevnik, A. (2016) *Jacques Lacan: Between Psychoanalysis and Politics*. London: Routledge.

United Nations Environment Programme. (2024) "Goal 12: Ensure Sustainable Consumption and Production Patterns." https://www.unep.org/topics/sustainable-development-goals/why-do-sustainable-development-goals-matter/goal-12-9 (last accessed May 2024)

United Nations Framework Convention on Climate Change. (1997) "Kyoto Protocol to the United Nations Framework Convention on Climate Change." https://unfccc.int/documents/2409 (last accessed May 2024).

United Nations Framework Convention on Climate Change. (2015) *Adoption of the Paris Agreement*, 21st Conference of the Parties. Paris: United Nations. https://unfccc.int/process-and-meetings/the-paris-agreement/the-paris-agreement (last accessed May 2024).

Vanhulst, J., and Beling, A. E. (2014) "Buen vivir: Emergent Discourse Within or Beyond Sustainable Development?" *Ecological Economics*, 101, 54–63.

Von Homeyer, I., Oberthür, S., and Dupont, C. (2022) "Implementing the European Green Deal During the Evolving Energy Crisis." *JCMS: Journal of Common Market Studies*, 60 (S1), 125–136.

Von Homeyer, I., Oberthür, S., and Jordan, A. (2021) "EU Climate and Energy Governance in Times of Crisis: Towards a New Agenda." *Journal of European Public Policy*, 28 (7), 959–979.

Wax, E., and Brzeziński, B. (2024) "Ursula von der Leyen Scraps Pesticide Reduction Bill, in Gift to Farmers." www.politico.eu/article/ursula-von-der-leyen-pesticide-reduction-bill-farmers/ (last accessed May 2024).

Weber, C. (1998) "Performative States." *Millennium – Journal of International Studies*, 27 (1), 77–95.

Weise, Z., and Guillot, L. (2024) "How the EU's Flagship Nature Law Became an Electoral Punching Bag." www.politico.eu/article/nature-restauration-law-european-parliament-election/ (last accessed May 2024).

Wodak, R. (2011) *The Discourse of Politics in Action*. Basingstoke: Palgrave Macmillan.

Wright, C. (2016) "Discourse and the Master's Lining: A Lacanian Critique of the Globalizing (Bio)Politics of the Diagnostic and Statistical Manual." In S. Tomšič and A. Zevnik (eds), *Jacques Lacan: Between Psychoanalysis and Politics*. London: Routledge, 131–145.

Wurzel, R., and Connelly, J. (2011) *The European Union as a Leader in International Climate Change Politics*. 1st ed. London: Routledge.

Zevnik, A. (2016) *Lacan, Deleuze and World Politics, Rethinking the Ontology of the Political Subject*. London: Routledge.

Zevnik, A. (2017) "A Return of the Repressed: Symptom, Fantasy and Campaigns for Justice for guantánamo Detainees Post-2010." *The British Journal of Politics and International Relations*, 20 (1), 206–222.

Žizek, S. (1993) *Tarrying With the Negative: Kant, Hegel, and the Critique of Ideology.* Durham, NC: Duke University Press.

Žižek, S. (2004) "The Structure of Domination Today: A Lacanian View." *Studies in East European Thought*, 56 (4), 383–403.

Žižek, S. (2006) *The Parallax View*. Cambridge, MA: MIT Press.

Žižek, S. (2013) *Less Than Nothing: Hegel and the Shadow of Dialectical Materialism*. London: Verso.

Appendix

Table A.1 EU side events attended at the COP23

UNFCCC COP23 Bonn (6–17 November 2017)	
06 November 2017	• LULUCF actions in European Member States (organised by DG Clima). • EU's climate diplomacy: Innovative approaches for a climate-resilient, low-carbon future (organised by Adelphi in cooperation with EEAS; German Federal Foreign Office). • Air quality and climate change (organised by DG RC).
07 November 2017	• Implementing the Paris Agreement: Energy Transition and Innovation and the role of the International climate governance (organised by L'Institut du développement durable et des relations internationals (IDDRI), Climate strategies (CS)). • Increasing action between now and 2020 (organised by CAN Europe). • JPI action: Actions in addressing shared challenges of climate change Providing climate knowledge to support societal innovation and transformation (organised by Joint programme initiative 'Connecting climate knowledge for Europe'; DG research and innovation, Centro Euromediterraneo sui cambiamenti climatici).
08 November 2017	• The EU steps up climate finance: the LIFE programme (organised by LIFE unit at EASME; DG Clima). • Just transition to a low-carbon economy: responding to political and ethical challenges of global climate change (organised by EESC, King's College London/Foundation for European Progressive Studies; Fondation Jean-Laurès.
10 November 2017	• LIFE BEEF carbon: the carbon mitigation action plan in beef production in France, Ireland, Italy, Spain (organised by IDELE French livestock institute INTERBEV). • Shared mobility for climate mitigation and big data (the real urban emission initiative (TRUE) (hosted by the International Transportation Forum and Institute for Transport and Development Policy (ITDP).

(*Continued*)

Table A.1 (Continued)

UNFCCC COP23 Bonn (6–17 November 2017)

| 11 November 2017 | • Transport thematic day Sustainable freight for a low-carbon transport system (moderated by Pat Cox, Former President, European Parliament).
• Visionary leadership for the transition to 1.5 temperature limit with clean technology (organised by Brahma Kumaris, The centre for Alternative technology). |

Table A.2 In-site observations conducted at the EU headquarters in Brussels

---

Fieldwork in Brussels, June—December 2018

13 June 2018	• *EU for Talanoa.* Stakeholder event organised by the EU Commission for sharing stories of success to meet the Paris Agreement goals.[1] The EU presented its story of success through the 2030 framework, and it also mentioned the EU is preparing its economic, environmental social analysis that will be published in November 2018 under the label 'EU long-term strategy'.
10–11 July 2018	• *The EU long-term vision for a clean, modern and competitive economy.* A two-day stakeholder consultation that brought together policy-makers and stakeholders from business, research, and civil society for a discussion on the forthcoming strategy. This stakeholder event preceded the opening of public consultations to take stock of all the various positions to be included in the long-term strategy.[2]
17 September 2018	• *Carbon Capture and Utilisation Technologies – Technological status, environmental impacts, and policy developments.* This stakeholder event on CCU gathered together academics, industry, and policy experts to discuss the current and future potential of CCU technologies, the efforts that need to be undertaken by industry and research, how these are linked to climate objectives, energy, impacts on other technologies (electrification, renewables, hydrogen).[3]
18–19 September 2018	• *Postgrowth Conference.* Multi-stakeholder gathering organised by ten Members of the European Parliament representing five political groups: Philippe Lamberts, Florent Marcellesi and Molly Scott-Cato (Greens/EFA), Alojz Peterle (EPP), Gerben-Jan Gerbrandy (ALDE), Marisa Matias and Helmut Scholz (GUE) and Guillaume Balas, Elly Schlein and Kathleen Van Brempt (S&D). The organisers cite that the aim was to 're-think future policies and discuss alternatives respecting the environment, human rights and viable economic development' (PostGrowth, 2018). But the panels organised aimed at discussing the myth of growth. It gathered academics, activists, and policymakers.[4]

(*Continued*)

Table A.2 (Continued)

Fieldwork in Brussels, June–December 2018

10–11 October 2018	• *EU Parliament ordinary meetings. ENVI committee meeting.*[5] ENVI is the Parliament committee dealing with climate. As its scope is much bigger than climate change, items on the agenda are not always climate-related. The first day was a voting day only, there was no parliamentary debate of the items on the agenda. On the second day a debate on air pollution was observed.
23 October 2018	• *CAN Europe General Assembly.*[6] Exceptional participation upon invitation. During the assembly, CAN Europe members shared views about the necessity to push the EU for more ambition (ambition is usually interpreted as higher targets). They also expressed some concerns over the not yet released Communication on long-term strategy as they are one of the stakeholders.
29 October 2018	• EESC event: *Facilitating access to climate finance for non-State actors.* The EESC is not an EU institution defined by the Maastricht Treaty (1992), but it in fact represents all civil society. As an institution, they work closely with the official EU institutions, most notably the Commission and the Parliament. The theme of this event was the problem of financial access to EU funds within the EU, including EU municipalities.
08 November 2018	• *Change the climate. For our health.* The event at the EU Parliament was attended upon invitation by the MEP Mr Pedicini (M5S). Another event that takes place by the initiative of a MEP and not by the EU Parliament as an institution. Because it was organised by an Italian MEP it focused more on air pollution, environment, and risks to health for mainly Italian municipalities.
12–16 November 2018	• *Critical raw materials week.* Stocktaking event organised by EU Commission's DG Grow. A five-day discussion between policy-makers and research institutes on the critical raw materials that are needed for the low-carbon transition and how these can be used in circular economy models. Experiences of 'sustainable mining' reported.[7]
19 November 2018	• *Black carbon & climate change in the European Arctic.* Organised by the Northern Dimension Institute,[8] the event focused on the neglected problem of black carbon emissions in the Artic coming mainly from Russia and affecting the Scandinavian countries. The event was advertised by the EU and included a keynote speech by DG Clima Yvon Slingenberg and the participation of DG environment at the panels.

(Continued)

Table A.2 (Continued)

Fieldwork in Brussels, June–December 2018

---

27 November 2018	• *EU Research and Innovation in our daily life.* An EU parliament/EU Commission joint event was organised at the EU Parliament by the former president of the EU Parliament Antonio Tajani, and the former Commissioner for Research, Science, and Innovation Carlos Moedas. This event brought together researchers and politicians to reflect on past and present achievements of EU-funded research and innovation. The themes of discussion were not entirely related to climate change but included health and wellbeing, a sustainable environment, a safe and secure society, and how to put innovation on the market.
27 November 2018	• *EU trade policy day.* EU stakeholder event on open and fair global trade, against protectionism. The event itself did not primarily have a climate focus but a session on "sustainable" trade was included. Events that are not climate-driven are useful to understand to what extent climate and environment are considered across all areas of the EU's policymaking.
28 November 2018	• *High level round table on decarbonisation of heating sector.* This event gathered leaders of the European sustainable and renewable heat industry associations that organise themselves under the DecarbHeat Initiative. The aim was to debate the contribution of the heating sector in homes and industry to the decarbonisation of the EU energy sector, future and existing pathways, technologies, and policies to make the heating sector more sustainable and ensure a cost-effective transition by 2050.
30 November 2018	• *Tackling premature obsolescence in Europe.* EESC event to address planned obsolescence to protect consumers and transition to a circular economy. In 2013, the EESC launched its Milestone opinion.[9] The EESC organised this event 5 years later to take stock of the action taken so far by the EU institutions, Member states and businesses regarding durability and reparability, planned obsolescence, impact on consumers and to debate new ways of action.
06–07 November 2018	• *Boosting circularity among SMEs.*[10] Networking event organised by EU Commission's DG environment under the premise that to enable the transition to a circular economy, economic actors and local authorities should all connect and support SMEs. The event included the presentation of best practices and ad hoc workshops. For example, I was assigned to a small company in Greece producing washing products and detergents. Various problems emerged on how to make it circular and possibly local in the life cycle of a product in Greece with current policies and legislations.

(Continued)

Table A.3 EU side events attended at the COP24

UNFCCC COP24 (3–14 December 2018)

---

10 December 2018	• EU2050 Covenant of Mayors (organised by the Commission, Global covenant of Mayors secretariat, Regional covenant of Mayors secretariats, Global and regional city networks).
11 December 2018	• Circular economy, the missing link in climate action (organised by the Netherlands ministry of infrastructure) • Energy day, Thematic session. Coal regions in transitions (organised by DG Energy).
12 December 2018	• EU2050 High level event on the strategy for long-term EU greenhouse gases emissions reduction (organised by the EU Commission). • EU2050 High level panel of the European decarbonisation pathways initiatives (Pierre Deschamps, DG research and innovation). • The role of carbon pricing in reaching Paris objective and EU's long-term decarbonisation strategy (organised by the European Parliament). • Achieving a net zero emissions energy system by 2050 (organised by GasNaturally).
13 December 2018	• EU2050 towards a thriving carbon neutral industry in Europe by 2050 (organised by Sustainable Process Industry through Resource and Energy Efficiency Public-Private Partnership; DG research and innovation; Carbon Market Watch (CMW). • EU2050 long-term strategy for decarbonisation of transport (Organised by Transport and Environment, DG clima, Wuppertal institute of climate, Environment and Energy, UN-Habitat ÖKO-Institut). • EU2050 Yes Europe, Youth engaging for a sustainable Europe (Organised by Alexandra Blin, Institute Jacques Delors, Paris).

Notes

1 As the Talanoa dialogue takes place within the United Nations Framework Convention on Climate Change (UNFCCC) this event presents the story of the EU to be shared in an international context, and the EU reassertion of its participation in the Paris Agreement.

2 As an interview with CAN Europe has clarified, it is not at these events that the Commission first hears such stakeholder opinions (Interviewee 1, 19/07/2018), but this is certainly one of the best ways to access the research field.

3 In 2017, DG Clima began a study to identify the potential of CCU technologies with the aim of assessing their readiness and assessing the EU regulatory set-up. The aim was to ensure the contribution of these technologies to climate mitigation. This study was conducted by Ramboll together with consortium partners Institute for Advanced Sustainability Studies (IASS Potsdam), Universität Kassel, Center for Environmental Systems Research, CE Delft, and IOM Law.

4 This means that it was not organised by the EU Parliament as an institution. The Parliament as the locus of democratic debate quite often hosts events of this type, but this does not mean that they are official and that they are relevant in the decision-making process.

5 ENVI stands for Committee on Environment, Public Health and Food Safety.

6 CAN Europe is one of the largest networks of environmental NGOs, they lobby the Commission and regularly participate in stakeholder events.

7 As one interviewee in DG energy will explain a few days later, this is another type of stakeholder event, in that it does not fit into a process, but it is only a stock-taking event (Interviewee 7, 13/11/2018).

8 A research institute in Finland, which receives funds from the EU.

9 The 2013 Milestone opinion (CCMI/112-EESC-2013–1904) by the EESC called for a ban on products with built-in defects designed on purpose to end a product's life prematurely so that a new product is bought. It also suggested that more information on the product's lifespan should be made available to consumers. Based on this opinion, the European Parliament voted back in July 2017 on a resolution to increase the products' lifetime and highlight benefits for consumers and companies.

10 The EU Commission adopted its Circular economy package in 2015 in order to promote the transition toward a circular economy.

Index

Note: Page numbers in *italics* indicate a figure and page numbers in **bold** indicate a table on the corresponding page.

For Product Safety Concerns and Information please contact our EU
representative GPSR@taylorandfrancis.com
Taylor & Francis Verlag GmbH, Kaufingerstraße 24, 80331 München, Germany

www.ingramcontent.com/pod-product-compliance
Ingram Content Group UK Ltd.
Pitfield, Milton Keynes, MK11 3LW, UK
UKHW022339100726
473146UK00010B/847